Philosophy and Ethics of AI

Volume 2

Series Editor

Aimee van Wynsberghe
Institut für Wissenschaft und Ethik
Bonn, Germany

This book series promotes informed debate and discussion of the ethical impact and philosophical consequences of the continued and changing use of artificial intelligence (AI) in a global society. It focuses on how current and future AI applications including robotics, virtual reality, augmented reality, and so on impact our conceptions of, as well as our ability to achieve, the good life and a good society. The series also explores broad philosophical problems in this context, including: the ontology of artificial, digital, virtual, and augmented entities; the epistemological consequences of the use of AI in health, politics, science, law, business, security and other social areas; and the changes of the interrelation between global society and nature due to these technologies. The book series welcomes contributions from a variety of disciplines and encourages inclusion of underrepresented voices in the global AI ethics debate.

Rosalie Waelen
Editor

Normative Relations of Recognition in the Age of AI

On Critical Theory and Technology

Springer

Editor
Rosalie Waelen
Institute for Science and Ethics
University of Bonn
Bonn, Nordrhein-Westfalen, Germany

ISSN 3059-3948 ISSN 3059-3956 (electronic)
Philosophy and Ethics of AI
ISBN 978-3-032-34027-6 ISBN 978-3-032-34028-3 (eBook)
https://doi.org/10.1007/978-3-032-34028-3

This Springer imprint is published by the registered company Springer Nature Switzerland AG
The registered company address is: Gewerbestrasse 11, 6330 Cham, Switzerland

Acknowledgements Although recognition and acknowledgment are distinct concepts, it would be odd for a book on recognition not to acknowledge those who inspired the project and helped it materialize. I want to express my deep gratitude to all the amazing authors who contributed a chapter to this edited volume: Christopher Senf, Leif H. Pedersen, Joel Anderson, Natalia Juchniewicz, Nynke van Uffelen, Fernando Secomandi, Florian Richter, Michał Wieczorek, Caroline Stitt, Christina Pinsdorf, and Arto Laitinen. I am very pleased that each of them enthusiastically accepted my invitation to write a book chapter on this topic, and it has been a tremendously gratifying experience for me to see their initial ideas turn into extensive texts. I hope that this book is only the beginning of our collaboration and research on the intersection of recognition theory and philosophy of technology.

The idea for this book emerged after Chris invited me to co-organize a panel on recognition in the age of AI for the 4TU.Ethics/ESDiT conference in the Netherlands, in 2024. Many of the contributors to this book were participants in the panel (Chris, Leif, Natalia, Michał, and myself) or members of the audience (Joel, Nynke, and Caroline). I'm indebted to Chris for initiating this panel and grateful to the panelists and audience members for their sincere interest in the topic. A special thanks also goes out to Michał, with whom I first explored the possibility of using recognition theory as a framework for the ethical analysis of AI applications, back in 2021 when we were both Ph.D. students. Furthermore, I thank Aimee van Wynsberghe (both in her role as series editor for the Springer book series on *Philosophy and Ethics of AI* and as director of the *Institute for Science and Ethics*) and all my colleagues in Bonn for our many conversations about the book. I thank my colleague

Christina in particular, for our friendly and fruitful disagreements about the (im)possibility of receiving recognition from AI systems. Finally, I am grateful to have had the opportunity to conduct a research visit at Macquarie University in Australia in 2025, where I had the honor and pleasure of working with Jean-Phillippe Deranty and meeting Heikki Ikäheimo. Our conversations about recognition and technology were a valuable source of inspiration for me when I was writing my own chapters for this book.

Bonn, 1 April 2026 Rosalie Waelen

Contents

Rethinking Recognition in the Age of AI

Rosalie Waelen

Abstract In the age of artificial intelligence (AI), the nature of social relations and power structures is changing fundamentally. AI and other recent technologies have brought about new means of communication, new bureaucratic tools, new educational practices, new career options, new types of companions, and more. While these technologies are potentially promising for promoting recognition in the normative and philosophically relevant sense, they can also introduce new forms or modes of social bias, inequality, domination, and self-distortion. If it is the case that relations of recognition are essential to individual self-development and societal justice—as many political philosophers and critical theorists claim—then the arrival of AI calls for renewed reflection on and theorization of recognition relations, their constitution, their challenges, and their implications. With the aim of rethinking recognition in the age of AI, this book brings together two fields of research that were once intimately related: critical social theory and philosophy of technology.

Keywords Recognition • Critical theory • Philosophy of technology • Artificial intelligence • Relational ethics • Social philosophy

R. Waelen (✉)
Institute for Science and Ethics, University of Bonn, Bonn, Germany
e-mail: waelen@iwe.uni-bonn.de

R. Waelen (ed.), *Normative Relations of Recognition in the Age of AI*, Philosophy and Ethics of AI 2, https://doi.org/10.1007/978-3-032-34028-3_1

1 Introduction

The central aim of this book is to explore the variety of ways in which artificial intelligence (AI) and other modern technological developments affect the constitution and functioning of recognition relations as well as the moral goods that derive from successful recognition relations. In doing so, the book not only perceives recognition theory as a valuable framework for normative analyses of emerging technologies, but also shows that the fundamental ways in which emerging technologies are changing the fabrics of society require critical theorists and political philosophers to revisit and possibly revise theories of recognition.

Technology's impact on society and social life was an important topic for first generation Frankfurt School critical theorists like Max Horkheimer, Theodor Adorno, and Herbert Marcuse. Deeply influenced by Karl Marx, the first generation of Frankfurt School thinkers critiqued technology (*Technik*) for its role in industrialization and modern capitalism, through which technology negated the objectives of enlightenment, produced a one-dimensional man, and enforced instrumental reasoning as a dominant mode of thinking. However, with later generations of critical theorists, the topic of technology faded into the background (Waelen and Bolte 2025).

It was third generation critical theorist Axel Honneth, a student of Habermas, who gave the Hegelian notion of 'the struggle for recognition' its central place in contemporary critical theory. One could say that he caused a paradigm shift in critical theory: moving beyond communication to recognition (Deranty 2009). Honneth first published his book *The struggle for recognition: The moral grammar of social conflict* in 1992 in German (which was translated to English in 1995 by Joel Anderson). Around the same time that Honneth published his seminal work on recognition, philosophers Charles Taylor and Nancy Fraser also wrote influential pieces arguing for placing recognition (back) at the center of political thought (Fraser 1995; Taylor 1992). A lot of scholarship on the notion of recognition followed, written by Honneth himself and many others (among other publications, see: van den Brink and Owen 2007; Deranty 2009; Fraser and Honneth 2003; Ikäheimo et al. 2021; Ikäheimo 2022). As a result, recognition theory has become an established and influential sub-domain of political philosophy and critical social theory (Iser 2019; Thompson 2006).

To the extent that this book is an attempt to strengthen the ties between critical social theory and the philosophy of technology, the focus on the topic of recognition is not an obvious one. The work of Axel Honneth and other recognition scholars is not explicitly about technology and seldom discussed within the philosophy of technology. The same goes for related fields like media studies and science and technology studies (STS). However, the chapters in this book will make clear that theories of recognition offer a unique and highly valuable source for understanding and critiquing technology's role in modern society, social life, and self-development. AI and other recent technologies are capable of fundamentally changing the means, content, scope, and impact of recognition relations, by altering the ways in which humans relate to the world around them, to other people, and to themselves. Modern technologies affect how we consume information and what topics and views we come across, how we structure societal institutions and how we get governed, how we communicate with one another and who we communicate with, how we perceive and treat other people, how we fall in love and who we fall in love with, and how we understand our own identity and find our place in the world. Admittedly, it is not exactly a new insight that technology changes how we relate to the world, to others, and to our self; the novelty of this book, in comparison to previous work on human-technology relations and AI's implications, lies in its focus on a specific type of relationality, namely, recognition relations.

Relations of recognition have a significant normative dimension, because someone's identity, self-worth, autonomy, social freedom, and integration in society, all rely on the recognition they receive externally (Anderson and Honneth 2009; Honneth 1992, 2014). Put differently, it is in interaction with other persons and societal institutions, that individuals achieve said normative goals (and possibly also in interaction with other lifeforms (see Pinsdorf 2016) or technological systems (see Chaps. 9, 10 and 11)). As a relational concept, recognition places the ontological conditions for human flourishing and social justice not with the individual, but in social relations and a society's institutional structures, which shape interpersonal social relations. A just society is one that enables social relations in which people can mutually receive and give recognition. Theories of recognition present an alternative to distributive theories of justice. Distributive theories emphasize that justice requires a fair distribution of material, social, or other goods, while recognition theories consider social relations to be the necessary foundation for justice. However, the exact relation between distributive justice and recognition justice has been widely debated. According to Honneth, whose work on recognition plays a central role in many chapters in this volume, distributive needs derive from recognition needs (Fraser and Honneth 2003). Fraser, on the other hand, argues that the two exist alongside each other, and are equally important to the goal of a just society (Fraser and Honneth 2003).

In this introductory chapter, I briefly discuss why and how the age of AI requires philosophers and critical theorists to rethink recognition. I do so through an overview of the chapters of this edited volume. The order of the chapters in this volume is based on four central themes. A first theme is *the struggle for recognition* (Sect. 2), which formed the basis for Hegel's and Honneth's thought on recognition. According to Hegel and Honneth, it is through struggles for recognition that social structures improve and moral progress is realized. AI changes struggles for recognition in many respects. AI offers new vehicles or means for social movements to fight for recognition, on the one hand, all the while it constitutes a new territory for misrecognition to fight over, on the other hand. The second theme focuses on the *normative implications of (mis)recognition* (Sect. 3). Implications that will be discussed have to do with identity, technology design, freedom, and alienation. A third theme looks into the *three spheres of recognition* (Sect. 4) outlined by Hegel, Honneth, and Taylor: (1) the sphere of personal relations where recognition is expressed in the form of love and care, (2) the political sphere where recognition takes the form of universal rights, and (3) the social sphere in which recognition implies relating to other members of society with solidarity and expressing esteem for their particular contributions to society. The fourth and final theme of the book has to do with *theoretical constraints* (Sect. 5). The focus of this theme is on the possibility of theorizing AI as more than a mere mediator of recognition, by analyzing whether AI can be considered an independent receiver or giver of recognition. In other words, the fourth part of the book explores under what conditions, if at all, human-AI relations can be recognition relations.

2 The Struggle for Recognition

Following the early work of G.W.F. Hegel, Honneth grounds his theory of recognition on the insight that social conflict can have a moral nature or "grammar" (Honneth 1992). The notion "*the struggle for recognition*" refers to the phenomenon that people unite and fight for societal change on the basis of shared experiences of injustice. To overcome experienced injustices, which stem from forms of misrecognition, the social structures that determine which demographics and what features receive recognition need to be transformed. Clearly, a theory of justice based on the notion of recognition and the transformative power of social conflict is indebted to critical theory's method of *immanent transcendence*. In line with the critical theory tradition, recognition theory does not present a preset image of how society should be, but rather aims to realize social and moral progress through a critique of the here and now. Social change is based on overcoming the normative failures experienced in the present. This also means that, instead of putting theory and moral principles first, recognition scholars (should) place the lived experience of the victims of injustice at the heart of political theory (I get back to this point in Chap. 11).

In Chap. 2 of this volume, Christopher Senf (University of Bergen, Norway) and Leif H. Pedersen (Roskilde University, Denmark) further discuss the role of the struggle for recognition in Honneth's work, in order to understand and critique the ways in which AI-based platforms exploit the fundamental human need for recognition. In Sect. 2 of Chap. 2, Senf and Pedersen explain that AI provides new means as well as new objects for recognition struggles. On the one hand, AI-driven social media platforms have proven to be a useful resource for victims of injustice and social movements. A vivid example of this is how the online use of *#MeToo* and *#BlackLivesMatter* enabled people from around the globe to unite their forces and fight for social change together. On the other hand, though, it has also become clear in recent years that AI-based tools such as recommender algorithms, personalization, and facial recognition, create new objects for social conflict. These tools and systems misrecognize people by reinforcing pre-existing social biases and power hierarchies, or creating new ones. Hence, while enabling people to fight for justice, AI also gives them new reasons to do so. Senf and Pedersen show that it is not only the issue of recognition itself that is ambivalent (Ikäheimo et al. 2021), but also its interplay with AI and social media platforms.

In Sect. 3 of Chap. 2, Senf and Pederson go on to argue that social media platforms manipulate and exploit the recognition needs of their users. Indeed, the entire business model of these platforms is based on the fact that humans have an innate desire for recognition and are therefore vulnerable to systems that promise to satisfy this hunger, but in fact feed it (see also Zuboff 2019). Dating apps, for instance, successfully sell premium options to their users when those users are most vulnerable—e.g. after having experienced rejection or heartbreak. Furthermore, dating apps and other AI-driven platforms know they can get people hooked on their services by fostering and exploiting a constant craving for more and renewed recognition. In other words, these platforms have managed to change social structures and social relations in such a way that people have become reliant on platforms' technological mediation to receive recognition from friends, colleagues, and romantic partners. By monetizing humanity's fundamental need for recognition, Senf and Pedersen write, businesses like social media platforms treat people as mere means to financial ends. This exploitation is morally objectionable, they argue, because there are four forms of misrecognition at the core of social media's exploitation of recognition needs: structural disrespect for informational self-determination, disregard for users' epistemic agency, a violation of democratic participation in recognition practices, and the objectification of recognition subjects.

Another important area of social life in which AI is changing recognition practices and gives rise to new types of struggles for recognition, is the workplace. In Chap. 3 of this volume, Joel Anderson (Utrecht University, the Netherlands) explains that the social sphere of labor traditionally has been a source for self-realization and social cohesion. Each worker contributes to society in their own way, and by doing so they gain a sense of self-esteem and social belonging. This is changing, however, due to the ways in which AI transforms workplaces and -practices. AI threatens the recognition realized through work by transforming the cooperative structures established within the sphere of work. Anderson analyzes three dimensions of work where AI has a transformative impact on the recognition relations previously realized through work: coordination, performance, and assessment. For each of these three issues, he discusses how AI changes what work is, how AI affects the autonomy of workers, and what AI means for the legitimacy and governance of workplaces. Concluding that the AI-mediated workplace is an important topic for contemporary critical theory, Anderson argues that recognition theory is distinctly capable of explaining how AI impoverishes the experience of workers as well as the social and democratic function of the workplace, even if the technology can enhance productivity and possibly even the quality of workers' outputs.

3 Ethical and Societal Implications

While many scholars are hesitant to consider AI as an independent actor in a recognition relation, the idea that technology can at least mediate recognition relations between human persons is now widely shared. As van Uffelen and Secomandi put it (in Chap. 5), AI and other technologies can materialize interpersonal relations of recognition. If AI can mediate relations of recognition, that means that recognition theory can complement AI ethics approaches that seek to understand the ethical and societal implications of AI applications. In recent years, various scholars have experimented with recognition theory as a frame of reference for analyzing the ethical and societal implications of technological innovation. For instance, Waelen and Wieczorek (2022) used Honneth's tripartite understanding of recognition to discuss the normative and psychological implications of gender biases in AI. Waelen (2022) showed how the philosophical concept of misrecognition can help to understand and critique the variety of ways in which facial recognition technologies can fail not only functionally, by inaccurately identifying or categorizing persons, but also morally, by failing to treat people as deserving of care, equal worth, and solidarity. Pedersen (2022) drew on Honneth's work on recognition to discuss the impact of mediatization on social change. Van Uffelen and Santos Ayllón (2025) used recognition theory to understand the injustices people experience in the context of the energy sector. Daus (2025) built on Honneth's and Fraser's work on recognition to describe the problem of conflicting values in AI systems used in the medical context. And Secomandi and van Amstel (2025), following Hegel, introduced a new notion in the context of design ethics: "*coerced recognition*". Coerced recognition, they explain, refers to "the phenomenological-dialectical process in which users are coerced into recognizing themselves as lesser beings in relation to designers when engaging with technology" (Secomandi and van Amstel 2025, p. 2).

Despite the agreement among these and other contemporary scholars that technology can mediate interpersonal recognition and, moreover, that recognition theory is a relevant resource for philosophers, ethicists, and others interested in the ethical and societal implications of emerging technologies, the relation between technology and recognition remains to be theorized. This edited volume serves that end. Chapters 4, 5 and 6 each help to deepen the understanding of the normative implications of technologically mediated recognition relations.

In Chap. 4, Natalia Juchniewicz (University of Warsaw, Poland) draws on recognition theory to explore the question: Can AI help shape individual and collective identities, in the way that intersubjective recognition relations do? In answering this question, Juchniewicz analyzes both the normative and epistemic dimensions of recognition. She explains that AI can affect identity not only by shaping a person's self-worth, in a normative sense, but also by shaping their self-understanding, in an epistemic sense. Although the constitutive and interpellative power of AI over identity is something to be wary of, Juchniewicz is cautiously optimistic that algorithmic categorizations (e.g. in the context of facial recognition tools) can also function as hermeneutic tools that enable people to perceive themselves in entirely new ways. In other words, she is open to the possibility that technologies function as an "other" through which we come to understand ourselves (as also argued by Gertz 2018 and myself in Chap. 11 of this book).

In addition to discussing AI's impact on identity formation, Juchniewicz argues that AI impacts what aspects of identity are considered important and whose identities get to be known in the first place. Drawing on Judith Butler's notion of *recognizability* (Butler 2021), Juchniewicz explains that algorithms (for instance on social media) determine what demographics and what characteristics are made recognizable to the outside world. This important and insightful critique of AI ties into the idea that AI and other emerging technologies are not only a mediator of interpersonal recognition or a means that assists in the exercise of struggles for recognition, but also a new source of misrecognition and therefore a new locus for recognition struggles (as also discussed in Chap. 2 by Senf and Pedersen).

In Chap. 5, Nynke van Uffelen (Radboud University, formerly TU Delft, the Netherlands) and Fernando Secomandi (TU Delft, the Netherlands) continue the discussion on recognition's ties to identity. Like Juchniewicz in Chap. 4, van Uffelen and Secomandi argue that, nowadays, interpersonal relations of recognition are constituted by technologies. If people can misrecognize each other through technologies, they reason, this implies that technologies can play a mediating role in identity formation and shaping a person's sense of self. On the basis of this insight, the authors analyze the implications of recognition-through-technology for design ethics and design practices. They argue that, since technologies are always designed, the potential for recognition and misrecognition through technology ought to be taken into account in the design of new technologies. Introducing the notion "*design for recognition*", the authors warn about four possible dangers that may occur when designing for recognition: (1) a distorted moral sensorium that compromises people's ability to feel and respond to experiences of misrecognition; (2) value trade-off between different types of injustices; (3) design is static but identity is not, which challenges people's ability to reject and transform recognition practices; and (4) designing for recognition comes with the risk of reifying certain identities in design and thereby, albeit unintentionally, suppressing others.

Furthermore, van Uffelen and Secomandi present recognition theory as an alternative approach to post-phenomenological analyses of human-technology relations.[1] They criticize post-phenomenology for treating individuals as isolated beings (Verbeek 2011), and argue that recognition theory is better able to account for the dialectical nature of subjectivity (see also Secomandi and van Amstel 2025). With the notion of 'misrecognition', recognition theory is also better able to explain the moral risks or pitfalls of technological mediation. Finally, the authors suggest that the notion of recognition is more suitable to include considerations of human-technology relations into design practices.

[1] For further discussions about the relation between recognition theory and post-phenomenology, see also Gertz (2018), Juchniewicz (2024), Secomandi and van Amstel (2025).

Chapter 6, written by Florian Richter (Catholic University of Eichstätt-Ingolstadt, Germany), focuses on the importance of freedom and how the conditions for freedom may be threatened by modern technologies. In his discussion of this issue, Richter contrasts Hegel's political philosophy with Plato's ideal state described in the seminal book *Republic* (Politeia). He explains that, while Plato's *Republic* is all about uniformity, Hegel's more modern and liberal approach values plurality and encourages the development of individual skills and interests. The risk we face today, however, is that emerging technologies serve Plato's ideal rather than a Hegelian one. Richter argues that emerging technologies, such as personalization algorithms, promote conformism and undermine people's development of critical and self-reflective cognitive capacities. As a result, personalization tools alienate people them from themselves.

4 Spheres of Recognition

Inspired by Hegel, Honneth (1992) distinguishes three 'spheres' or 'axes' of recognition: *love*, *respect*, and *social esteem*. Each of these three spheres of recognition has its own distinct features as well as its own value for a person's relation-to-self. Relations of love imply genuine care for an individual's feelings, desires, and well-being. Respect pertains to generalized relations between members of society, which are institutionalized through rights. Social esteem, also referred to as solidarity, exists in social relations in which an individual receives recognition not for their universal worth, but their unique individual contributions to society. Since each sphere of recognition is important in its own right, a successful society needs to provide a setting in which members of society have access to all three types of social bonds. However, as Honneth himself acknowledges, this tripartite distinction is not set in stone:

> we must always reflexively examine the boundaries that have been established between the domains of the different recognition principles, since we can never rule out the suspicion that the existing division of labor between the moral spheres impairs the opportunities for individual identity-formation (Fraser and Honneth 2003, p. 189)

Given that recognition can take different forms, it is worthwhile considering how AI may impact some spheres of recognition more strongly or differently than others. AI may successfully mediate one type of recognition while forming a hurdle to another, or introduce new modes of recognition while making others obsolete. To this end, some of the contributions to this book zoom in on one particular type of recognition.

Chapter 7, written by Michał Wieczorek (University College Dublin, Ireland), aims to deepen the understanding of recognition as love between children and adults. He explains that children cannot always voice their thoughts and feelings with the level of sophistication that we expect from adult persons. Therefore, he argues, recognition relations with children require adequate interpretative skills from the adults who interact with them. Through his discussion of recognition relations with children, Wieczorek seeks to understand and critique the increasingly dominant role that algorithmic technologies play in children's development. He is sceptical that the algorithmic technologies that children increasingly interact with can successfully simulate the aspects of social interaction with others that children need to experience in order to develop into sociable adults. Such aspects include acquiring knowledge of the interaction partner, establishing trust, and forging emotional connections. Furthermore, successful recognition of children not only depend on the child's ability to express their needs and wants, but also on the hermeneutic skills of the giver of recognition—skills which algorithmic technologies lack, according to Wieczorek. He therefore concludes that using modern technology for children's socialization should be approached carefully and critically.

In Chap. 8, Caroline Stitt (AI + Planetary Justice Alliance, Germany) discusses the ways in which respect is and should be expressed in AI policy. Due to the ubiquitousness of AI in nearly every aspect of people's lives, policies that guide the development and use of AI applications are urgently needed to ensure that people's fundamental rights and values are protected. The notion of respect offers a unique explanatory frame for the importance of AI policy. Honneth's notion of recognition as respect, related to Taylor's politics of universalism (1992), entails that people are treated as bearers of rights who have equal moral worth as their fellow members of society. The antonym, disrespect, then refers to experiences of being overlooked or treated as less than equal. Through an in-depth critical analysis of the European Union's AI Act, Stitt argues that an intent to respect is present in the Act's risk-based approach, human oversight provisions, transparency obligations, and compliance measures. In practice, however, the EU's AI policy falls short in protecting and promoting the autonomy and self-development of AI users and EU citizens, because of weak enforcement mechanisms, a strong reliance on industry self-regulation, and significant loopholes in the legislation. Additionally, Stitt asks the critical questions: Who should be protected and respected by AI legislation? Even if the AI Act is a European piece of legislation, the development of

AI by European companies and use of AI by European citizens affects the well-being and autonomy of others too—like migrants and refugees, AI workers, and those affected by the environmental costs of AI—who are equally deserving of respect.

Although no chapter in this volume is dedicated specifically to the third strand of recognition—social esteem and solidarity—it is worthwhile pointing to an article by Nolen Gertz (2018) on relations of solidarity between soldiers and the robots that assist them. As one of the first and only scholars to connect recognition theory with the philosophy of technology, Gertz argues that users can experience solidarity with robots when the latter fulfill social roles that are functionally equivalent to the social roles of the user—as is the case in the context of soldiers and military robots. Following Hegel's analysis of the struggle for recognition, Gertz explains that it is through encounters with others that one becomes conscious of the self. What matters for Hegel is only that the other *appears* to be conscious, he continues. Arguably, AI systems and robots do *appear* to be conscious, even if they cannot be ascribed true consciousness. So, he concludes, when humans interact with technologies that are functionally equivalent to them, their self-understanding can be improved by the human-technology relation. Because recognizing robots as others can improve a person's relation-to-self, functionally equivalent robots should be understood as potential objects of solidarity and members of the moral community.

Cappuccio et al. (2020) also argued that humans can have recognition relations with robots and, moreover, that robots can be the recipient, i.e., moral patient, in such a recognition relation. However, their reasoning differs from Gertz's. Cappuccio and colleagues argue that social robots should be conceived as worthy of moral recognition, because humans can cultivate social virtues in their quasi-social relations with robots. Not treating human-robot relations as recognition relations is then a dangerous breeding ground for the cultivation of social vices, i.e., social and moral deskilling. In Chap. 9 of this chapter, Pinsdorf continues these discussions on the possibility of AI systems as a recipient of recognition. Unlike Gertz (2018) and Cappuccio et al. (2020), she concludes that technology can not be a recipient of recognition in a recognition relation.

5 Theoretical Constraints

The chapters outlined so far have mostly focused on AI as something that can change, mediate, or otherwise affect relations of recognition between human persons (i.e. *interpersonal recognition*). The final part of the volume looks beyond the impact of AI and digitalization on interpersonal recognition, by asking whether direct relations of recognition between humans and the technologies they use are possible too. Chapters 9, 10 and 11 attempt to address the question: Can humans have recognition relations with technologies like AI systems or robots? Answering this question requires revisiting the very fundaments of recognition theory, such as the ontological features that actors in recognition relations ought to possess and the moral value of relationality as such.

How to call direct recognition relations between humans and technology (as opposed to technologically mediated recognition) remains an open question. The different authors in this book have proposed various terms: "synthetic recognition" (Anderson, Chap. 3), "simulated recognition" (Laitinen, Chap. 10), "pseudo-recognition" (Pinsdorf, Chap. 9), "algorithmic recognition" (Senf and Pedersen, Chap. 2), and "artificial recognition" (Waelen, Chap. 11). Part of rethinking recognition in the age of AI is then also to establish new conceptual tools for making sense of the new modes and types of recognition that AI brings about.

In Chap. 9, Christina Pinsdorf (University of Bonn, Germany) introduces a novel typology of recognition relations, which she previously developed to capture moral relations between humans and non-human lifeforms (Pinsdorf 2016). Pinsdorf's typology is based on a distinction between symmetry and reciprocity in recognition relations. Symmetry, she argues, refers to the moral status of the agents and patients in a recognition relation, while reciprocity entails that the recognition relation is realized mutually. Recognition scholars usually conflate these two terms, but Pinsdorf argues that it is important to separate them. Doing so allows her to theorize recognition relations with non-human lifeforms. On the basis of her typology of recognition relations, Pinsdorf then systematically analyzes whether AI systems can be a moral agent (i.e., a giver of recognition) or a moral patient (i.e., a recipient of recognition) in a recognition relation. She concludes that AI cannot be a moral agent because it is not capable of moral reasoning or moral motivation and, therefore, incapable of giving *genuine* recognition. AI also cannot be a moral patient, according to Pinsdorf, because they do not have any innate qualities that make them objects of our moral duties.

In Chap. 10, Arto Laitinen (Tampere University, Finland) analyzes the relations between humans and robots in light of the human desire for recognition. Like Pinsdorf, he revisits the requirement of mutuality in recognition relations and the capacities that moral agents and patients in recognition relations ought to have. Though the spirit of this chapter is similar to the previous one, Laitinen takes quite a different approach to assessing the possibility of recognition relations between humans and robots. According to Laitinen, the fundamental philosophical questions that arise in this context can be addressed in roughly two manners: from the perspective of a scientific worldview and from a common-sense phenomenological perspective. He goes on to argue that the manifest common-sense worldview prevails with regards to recognition, as this view is presupposed in the practice of mutually recognizing each other. "Recognition must be recognized", he argues. Following the common-sense approach, relations of recognition between humans and robots are not possible because robots lack the relevant capacities (e.g. sentience, suffering, care) to speak of such relations, Laitinen concludes. Similar to Pinsdorf, he emphasizes the fundamental importance of personhood and genuineness for recognition.

However, despite both denying the possibility of recognition relations between humans and technology, Pinsdorf (Chap. 9) and Laitinen (Chap. 10) acknowledge that people may nevertheless *experience* their interactions with AI systems and robots as recognition. Using the term "*pseudo-recognition*", Pinsdorf explains that these systems are designed for the very purpose of creating such an illusion. Laitinen, in turn, speaks of "*simulated recognition*". Pinsdorf and Laitinen both end their chapters wondering what the moral value of pseudo-recognition or simulated recognition could be. Pinsdorf asks "can simulation suffice?" (Chap. 9) and, in a similar vein, Laitinen poses the question "could simulated recognition be enough?" (Chap. 10).[2] While Pinsdorf firmly defends the importance of philosophical and conceptual clarity about the ontological conditions for moral relations, Laitinen carefully opens the door for the possibility that simulated recognition may affect people's relation-to-self in relevant ways, by discussing how machines can send normatively laden messages that may lead to psychological harm.

Chapter 11, the final chapter of the book, is written by myself, Rosalie Waelen (University of Bonn, Germany). I agree with my colleagues in Chaps. 9 and 10 that the kind of recognition that AI systems can display does not meet the traditional criteria for genuine interpersonal recognition. However,

[2] Please note that both authors wrote their contributions simultaneously and without having read the work of the other. The similarities in their research questions are therefore merely coincidental.

I emphasize the fact that the present social reality is such that humans increasingly often have social interactions and social bonds with AI systems—or as I like to call them: "*artificial others*". Existing literature on the topic of recognition is limited in its ability to make sense of the nature and potential normative impact of this emerging reality. Therefore, I argue, it is important to develop theoretical tools that enable us to understand and critique this emerging phenomenon. Social bonds between humans and technology are a new social fact, and it is the task of social critique to understand the social reality we live in, and to lay bare and overcome its moral limitations. If current theories of recognition cannot capture the kind of (mis)recognition that people can experience in their interactions with AI systems, new theoretical work on the topic is needed. In my view, recognition theory ought to place practical experiences of injustice before theoretical limitations. If theory and practice do not align, it is the theory that should be rejected, not the practical experiences of victims of injustice.

So, in contrast to the previous chapters, Chap. 11 focuses on the fact that artificial (mis)recognition, despite being ungenuine, is nevertheless experienced as (mis)recognition. Through a discussion of three major theoretical hurdles—reciprocity, genuineness, and functionality—I try to lay the groundwork for a theory of recognition from artificial others. With regards to reciprocity, I argue that whether or not a recognition relation is reciprocal depends entirely on the recipient of recognition. In contrast to Laitinen (Chap. 10) and others who suggest that the giver of recognition not only needs to recognize the other but also themselves, I propose that reciprocity merely entails that the recipient recognizes the giver as an entity capable of giving recognition.[3] Furthermore, I critique existing literature on recognition for failing to adequately explain why genuineness is not just important (because it is obvious that it is), but also a *necessary* criterium for recognition. Through a discussion of Ikäheimo's and Laitinen's (2007) distinction between recognitive acts and recognitive attitudes, I argue that recognitive acts alone, without genuine recognitive attitudes, can have normatively relevant psychological implications on the part of the recipient of recognition. Hence, I argue that recognitive acts by artificial others can, in practice, function as recognition. On the basis of this argument, I then propose a new conceptual framework for recognition relations, that distinguishes *genuine recognition* from merely *functional recognition* (i.e., perceived recognition). The category of functional

[3] Laitinen (Chap. 10) argues that a recognition relation in which A mutually recognizes B implies four different relations: A recognizes A, A recognizes B, B recognizes A, and recognizes B. In Chap. 11, I argue instead that only the former three relations are needed to say that A reciprocally recognizes B. This means that B does not need to have self-consciousness or a reflective relation-to-self.

recognition not only captures experiences of (mis)recognition in interactions with AI systems or robots, but can also make sense of other types of experiences involving recognitive acts that were not motivated by genuine recognitive attitudes—such as ideological acts of recognition, cultural habits, or interactions between humans and their pets.

In conclusion, one way of reading this edited volume is to perceive each chapter as another argument in support of the thesis that we need to revisit recognition theory in the age of AI. In the context of experiences of recognitive interactions between humans and AI systems, we not only need to revisit recognition theory, but also revise it.

References

Anderson, J., and A. Honneth. 2009. Autonomy, Vulnerability, Recognition, and Justice. In *Autonomy and the Challenges to Liberalism*, edited by J. Christman and J. Anderson, 127–140. Cambridge University Press.

Butler, J. 2021. Recognition and the Social Bond: A Response to Axel Honneth. In *Recognition and Ambivalence*, edited by H. Ikäheimo, K. Lepold, and T. Stahl, 31–53. Columbia University Press.

Cappuccio, M. L., A. Peeters, and W. McDonald. 2020. Sympathy for Dolores: Moral Consideration for Robots Based on Virtue and Recognition. *Philosophy and Technology* 33. https://doi.org/10.1007/s13347-019-0341-y.

Daus, Z. 2025. Distribution, Recognition, and Just Medical AI. *Philosophy and Technology* 38 (1): https://doi.org/10.1007/s13347-025-00842-3.

Deranty, J. P. 2009. *Beyond Communication: A Critical Study of Axel Honneth's Social Philosophy*. Leiden: Brill.

Fraser, N. 1995. Recognition or Redistribution? A Critical Reading of Iris Young's 'Justice and the Politics of Difference'. *The Journal of Political Philosophy* 3 (2): https://doi.org/10.1111/j.1467-9760.1995.tb00033.x.

Fraser, N., and A. Honneth. 2003. *Redistribution of Recognition? A Political-Philosophical Exchange*. Verso.

Gertz, N. 2018. Hegel, the Struggle for Recognition, and Robots. *Techné: Research in Philosophy and Technology* 22 (2):

Honneth, A. 1992/1995. *The Struggle for Recognition. The Moral Grammar of Social Conflicts*, translated by Joel Anderson. MIT Press.

Honneth, A. 2014. *Freedom's Right: The Social Foundations of Democratic Life*. Columbia University Press.

Ikäheimo, H. 2022. *Recognition and the Human Life-Form: Beyond Identity and Difference*. Routledge.

Ikäheimo, H. and Laitinen, A. (2007). Analyzing Recognition: Identification, Acknowledgement, and Recognitive Attitudes towards Persons. In Recognition and Power. Edited by B. van den Brink and D. Owen, 33–56. Cambridge: Cambridge University Press.

Ikäheimo, H., K. Lepold, and T. Stahl, eds. 2021. *Recognition and Ambivalence.* Columbia University Press.

Iser, M. 2019. Recognition. In *Stanford Encyclopedia of Philosophy*, edited by E. N. Zalta. Stanford University. https://plato.stanford.edu/entries/recognition/.

Juchniewicz, N. 2024. Towards (Unilateral) Recognition of "the Technological Other"—Vulnerability, Resistance and Adequate Regard. *Ethics in Progress* 15 (2):

Pedersen, L. H. 2022. Furries, Freestylers, and the Engine of Social Change: The Struggle for Recognition in a Mediatized World. *Communications* 47 (4): https://doi.org/10.1515/commun-2022-0047.

Pinsdorf, C. 2016. *Lebensformen und Anerkennungsverhältnisse: Zur Ethik der belebten Natur.* De Gruyter.

Secomandi, F., and F. M. C. van Amstel. 2025. Coerced Recognition Through the Service Interface: A Design Ethics Framework for Revealing User Oppression. *Journal of Human-Technology Relations* 3. https://doi.org/10.59490/jhtr.2025.3.7745.

Taylor, C. 1992. The Politics of Recognition. In *Multiculturalism: Examining the Politics of Recognition*, edited by A. Gutmann. Princeton University Press.

Thompson, S. 2006. *The Political Theory of Recognition. A Critical Introduction.* Polity Press.

van den Brink, B., and D. Owen, eds. 2007. *Recognition and Power.* Cambridge University Press.

van Uffelen, N., and L. M. Santos Ayllón. 2025. Categorizing Experiences of Misrecognition in Energy Contexts: A Recognition Justice Typology. *Applied Energy* 389. https://doi.org/10.1016/j.apenergy.2025.125730.

Verbeek, P. P. 2011. *Moralizing Technology. Understanding and Designing the Morality of Things.* University of Chicago Press.

Waelen, R. 2022. The Struggle for Recognition in the Age of Facial Recognition Technology. *AI and Ethics* 3 (1): https://doi.org/10.1007/s43681-022-00146-8.

Waelen, R., and L. Bolte. 2025. Artificial Intelligence, Critical Theory, and the Concept of Progress. In *A Companion to Applied Philosophy of AI*, edited by M. Hähnel and R. Müller. Wiley.

Waelen, R., and M. Wieczorek. 2022. The Struggle for AI's Recognition: Understanding the Normative Implications of Gender Bias in AI with Honneth's Theory of Recognition. *Philosophy and Technology* 35 (53): https://doi.org/10.1007/s13347-022-00548-w.

Zuboff, S. (2019). The Age of Surveillance Capitalism. The Fight for a Human Future at the New Frontier of Power. Profile Books.

Algorithmic Exploitation and Recognition Struggles

Christopher Senf and Leif H. Pedersen

Abstract This chapter examines how AI systems give rise to new objects and means for recognition struggles—from biased facial recognition to engagement algorithms that subvert protest goals, to digital institutional systems that rank individuals. The focus is on how AI-powered social media platforms and their underlying algorithms strategically exploit fundamental human needs for recognition, which drive online engagement. Growing concern centers on how platform companies deploy advanced AI systems to capitalize on our social hunger for bonding, reward, and the bestowal of esteem, converting vital recognition needs—that is, our need for affirmation by others as the basis of our social life—into vulnerabilities to be monetized. However, recognition theorists examining contemporary protest movements and online activism have largely overlooked these effects of AI systems on struggles for recognition. We propose a framework for understanding the AI-enabled exploitation of recognition needs through algorithmic recognition mechanisms, debate why such exploitation is objectionable, and theorize how it may give rise to new forms of recognition struggles that demand urgent attention.

C. Senf (✉)
Department of Philosophy, University of Bergen, Bergen, Norway
e-mail: Christopher.senf@uib.no

L. H. Pedersen
Department of Communications and Arts, Roskilde University, Trekoner, Denmark
e-mail: hemming@ruc.dk

R. Waelen (ed.), *Normative Relations of Recognition in the Age of AI*, Philosophy and Ethics of AI 2, https://doi.org/10.1007/978-3-032-34028-3_2

Keywords Recognition theory • Artificial intelligence • Algorithmic exploitation • Algorithmic recognition • Misrecognition • Synthetic recognition • Social media platforms • Digital activism • Protest movements

1 Introduction

This section briefly introduces the theoretical foundations of recognition struggle theory, examining how collective experiences of misrecognition generate organized resistance and how these dynamics are being transformed by emerging forms of technological mediation in the age of AI. Recognition struggle theory draws its appeal from the notion that individuals depend on the affirmation of others (persons and institutions) for a positive relation to themselves, and that harm to affirmation can incite actors to rise up (Honneth [1992] 2012,[1] 320; Iser 2019; Ikäheimo et al. 2021, 1). Axel Honneth, along with Charles Taylor (1994) and Nancy Fraser (2017), offers one of the most influential theories of recognition (Thompson 2006). In Honneth's framework, love, respect, and esteem are institutionalized forms of affirmation—serving as enabling conditions for a free life for all—while acts of disregard are understood as threats to that free life. Crucial to this chapter is Honneth's argument that movement protests can be traced back to collective experiences of injustice, stemming from violated expectations of affirmation and disturbed social relationships—i.e., denials of recognition, such as the denial of rights. These experiences equip agents with motivating reasons to resist such threats (Renault 2019, 33ff) and to act jointly against perceived unjust conditions in pursuit of future recognition (Honneth 2012, 113).

On a conceptual level, recognition in the Honnethian sense is best understood as a reciprocal relationship between subjects—each relating to the other and mutually shaping themselves in light of the other's claims—rather than as a one-sided interaction between an active recognizer and a passive recognizee (Honneth 2018, 187 ff). Recognition involves two key dimensions. The first is cognitive—*re-Cognition*—the act of perceiving, identifying, or acknowledging the presence or features of another. The second is practical—*Re-cognition*—in the sense of valuing, appreciating, esteeming, or respecting the other (Iser 2008, 165 f.; Pollmann 2008, 28 f.).[2]

---

[1] This chapter draws heavily on *The struggle for recognition: The Moral Grammar of Social Conflicts*, by Axel Honneth (2012).

[2] Recognition is an "evaluative attitude" (Zurn 2015, 25), in which one subject perceives and affirms another's personhood, opinion, or actions within a shared relational dynamic. An act of recognition requires knowledge that someone or something possesses valuable features (e.g., emotional vulnerability),

Recognition theory offers a compelling lens for examining protest movements. Acts of recognition are crucial for individuals to develop autonomy and participate fully in social life (Deranty 2009, 272 f.). Conversely, the persistent denial of recognition undermines these capacities, sparking social critique, resistance, and eventually raising claims for acknowledgment (Honneth 2012, 219 ff.). Recognition theory holds that movements emerge when people experience a deep dissonance between widely endorsed norms and their insufficient realization in everyday life. Rooted in shared experiences of injustice, these protests reflect deeper societal tensions—in fact, intra-societal struggles between those who feel misrecognized and those they perceive as misrecognizers. When conventional avenues of redress are blocked, affected individuals form social networks and build collective agency through solidarity and shared convictions. The strength of this framework lies in its capacity to translate moral indignation into an analytical structure that identifies the underlying harms (Zurn 2015, 57).

On this view, protest is not merely political expression but an expression of collective moral injury, and its demands articulate a distinct "moral grammar" (Honneth 2010, 265). A protester may be seen as one that suffers moral wrongs and joins with others to raise claims with those who can recognize such claims, for instance, by granting rights, raising salaries, defining new cultural practices, or improving political inclusion. Protest movements are characterized by long-term campaigns of joint claim-making and protest action in public—external to formal state channels—over contested objects (e.g., wages, rights, and liberties), through non-institutionalized political means, and with regard to targeted agents (e.g., corporations, or government institutions) from whom they seek recognition of their claims (Tilly and Tarrow 2015, 145 ff). Recognition theory provides philosophical guidance to these morally led formations and seeks to anchor its assessments in the experiences of those who suffer injustice, aiming to contribute to movements' self-understanding rather than to impose external judgments (Honneth 2012, 113; Haslanger 2012, 22). It offers both an explanatory approach to make protest activities intelligible by examining the motivating reasons and moral

an acceptance that these features are valid or true, and a practical willingness to acknowledge them (Ikäheimo and Laitinen 2007, 34–37). For recognition to be complete, this attitude must be expressed—through words, gestures, actions, or institutional forms like legal rights—in a manner that allows the other to understand it as meaningful and responsive acknowledgment. Recognition thus becomes a mediated process in which a recognized subject can identify and respond to the affirming stance of another recognizing subject (Zurn 2015, 52). Honneth's concept extends beyond symbolic affirmation, spanning from Kantian respect for equal status—institutional forms that affirm moral autonomy—to esteem for individual contributions, and, ultimately, to care that affirms the emotional particularity and vulnerability of persons.

justifications for resistance, and an evaluative approach for assessing movements' normative claims against recognition theory's principles of justice. This chapter emphasizes the explanatory approach to understand how AI systems transform the conditions under which recognition struggles emerge.

Today, these dynamics of collective resistance and claim-making unfold within fundamentally transformed conditions. Recognition relations are increasingly mediated by technological processes of communication; media scholars term this the "mediatization" of social processes (Couldry and Hepp 2017, 3–4).[3] In fact, emerging technologies have created new infrastructures through which recognition struggles emerge and develop, while simultaneously introducing novel forms of misrecognition that generate their own resistance movements. In particular, capable computational systems—designed to perform tasks traditionally requiring human cognitive capacities such as perception, classification, decision-making, or language use—appear to take on roles that resemble those of recognizing subjects. These systems operate as what we might call "algorithms of recognition" (Susskind 2018, 275 f): AI technologies that simulate key aspects of recognition by identifying, sorting, and evaluating human traits and behaviors.

While recent advances in generative AI raise important questions about relationality and agency, this chapter examines more established AI systems and their specific applications—e.g., recommendation algorithms, personalization systems, and computer vision technologies.[4] Our analysis concentrates particularly on machine-learning-enabled recommender systems within social media platforms, which illustrate how algorithmic pattern recognition and learning capabilities transform recognition practices by continuously adapting to user behavior and creating algorithmic validation mechanisms designed to exploit human needs for acknowledgment. These systems—including content curation algorithms on Facebook, TikTok's "For You" page, and Instagram's feed algorithms—represent AI technologies that learn user behavior patterns and adapt their responses to maximize engagement. Throughout this analysis, we use 'AI systems' to refer to this broader category of technologies that employ machine learning capabilities to identify, predict, and respond to human behavior, exploring how these technologies emerge as both objects of struggle and as means for recognition struggles. Our primary concern lies

[3] Although such interactions are almost by definition characterized by a certain 'mediation'—either in terms of interpersonal communication (Honneth 2001, 119–120) or in terms of how humans are represented in and by institutions such as the state or the public sphere—both vertical and horizontal recognition relations (Ikäheimo 2017) now face qualitatively different forms of technological mediation.

[4] Broader questions of mutuality and agency are addressed elsewhere in this volume, particularly in Arto Laitinen's chapter.

with how these systems exploit recognition needs through personalized content recommendation and algorithmic validation mechanisms.

2 Recognition Struggles in the Age of AI

This section examines how AI systems fundamentally transform recognition struggles, analyzing through concrete cases—from Amazon Rekognition controversies to Black Lives Matter hashtag campaigns—what activists struggle against and the tools they must use to pursue social change. Building on the recognition theory framework outlined above, this analysis demonstrates two primary ways that AI technologies reconfigure these conflict dynamics: first, as objects of struggle themselves, where algorithmic misrecognition generates new forms of resistance demanding proper recognition and privacy rights; and second, as means of struggle that activists must navigate, where AI-powered social media platforms simultaneously enable protest mobilization while constraining collective action through the competitive logics that underlie engagement algorithms.

2.1 AI Systems as Objects of Struggle

A striking example of how AI systems are reshaping the emerging landscape of conflictive recognition relations is Amazon Rekognition—a computer vision tool developed by Amazon Web Services (AWS). Marketed as an easy-to-use and scalable solution for facial recognition and object detection, Rekognition enables the identification, verification, or tracking of individuals by comparing facial images against large databases (AWS 2025b). Beyond commercial use, AWS promotes the system for public and private safety purposes, including content moderation and identity verification. Importantly, Amazon Rekognition has been adopted by U.S. law enforcement agencies—including local police forces, the FBI, and Immigration and Customs Enforcement (ICE)—as part of automated identification processes.

In 2018, the American Civil Liberties Union (ACLU) tested Amazon Rekognition's accuracy by comparing photos of 535 members of Congress with 25,000 mugshots (Snow 2018). The results were alarming: the system falsely identified 28 members of Congress as criminals, with nearly 40% being people of color, though people of color comprised only 20% of Congress. These findings not only called into question the credibility of a technology marketed as offering "highly accurate facial analysis" (AWS 2025a) but also underscored serious concerns about racial bias embedded in AI systems. In response, the ACLU launched a campaign urging law enforcement agencies to halt their use of Rekognition, citing risks to privacy, civil liberties, and racial justice. The public pressure yielded results: in 2020, Amazon announced a one-year moratorium on police use of the tool—a suspension that was later extended indefinitely (ACLU 2020).

This story is significant for several reasons. First, it illustrates how AI systems are increasingly embedded in institutional practices of automated identification and law enforcement, shaping how persons are categorized, ranked, surveilled, and ultimately perceived and treated within these settings.[5] Second, it demonstrates that these systems are far from infallible: they can misclassify, exclude, and distort individuals, often as a result of being trained on biased datasets—thereby enacting what might be called *algorithmic misrecognition*. Third—and most importantly for this chapter—it highlights how such misrecognition can spark new struggles for recognition in the age of AI.

For instance, Waelen (2022) argues that struggles like these should be understood as demands for proper recognition, or the cessation of misrecognition, by AI systems. She shows that algorithmic misrecognition, particularly when facial recognition tools fail to accurately identify people of color due to racially biased training data, is not merely a technical flaw (cf., Bartoletti 2020; Noble 2018; O'Neil 2022). Rather, such failures mirror and reinforce existing social hierarchies, can produce real harm, and can be experienced as moral injury. When people are misclassified, excluded, or rendered invisible by intelligent digital systems—on the basis of race, class, or gender—this misrecognition constitutes a denial of moral equality. For those affected, it can generate feelings of being unseen, devalued, or treated as less than fully human, fueling resistance and demands for changes in how these technologies are developed and deployed. Civil society organizations such as the Algorithmic Justice League (AJL 2025) have emerged as leading voices in this growing movement.[6]

[5] In *Future Politics* (2018), Susskind recounts a 2016 case from New Zealand in which a man of Asian descent had his online passport application rejected because the automated system mistakenly determined that his eyes were closed in the submitted photograph (Susskind 2018, 276).

[6] The Algorithmic Justice League (AJL) actively mobilizes against AI-induced injustices, particularly those stemming from facial recognition technology. The AJL draws attention to how such systems are frequently trained on biased or exclusionary datasets—an outcome often rooted in the lack of diversity among developers and a broader culture of ignorance within the tech industry. These biases result in concrete harms for individuals and communities that are systemically misrepresented or ignored. The AJL advocates for bans on both governmental and commercial uses of AI that undermine the equal recognition of all human beings.

Beyond demands for proper recognition, AI-related struggles increasingly extend into the domain of privacy—specifically, to collective efforts aimed at remaining entirely undetected by such technologies in public, private, and digital spaces. During the Los Angeles anti-ICE protests in June 2025, several autonomous Waymo vehicles were set on fire. Activists argued that the vehicles—equipped with advanced computer vision systems and multiple cameras to analyze their surroundings—also operated as surveillance tools, potentially funneling data to law enforcement forces and threatening the anonymity of demonstrators (Hawkins 2025).[7] In terms of recognition theory, efforts to remain entirely undetected by AI systems can be seen as struggles for proper respect of the right to exercise self-determination over personal data—specifically, recognizing epistemic agency and control over one's digital data as fundamental conditions for personal autonomy (cf., Laitinen and Sahlgren 2021; Couldry and Mejias 2019, Chap. 5).

These cases illustrate that AI is emerging as a new object of struggle due to its impact on social life. AI systems are not mere technical tools prone to occasional errors. Rather, they actively mediate how individuals and groups are perceived, categorized, and valued within society. (Mis)recognition by these systems—understood here as cognitive or technical processes of algorithmically (mis)spotting or (mis)identifying features and patterns (see chapter by N. Juchniewicz, this volume) and drawing conclusions against preset metrics—can lay the groundwork for inappropriate treatment and experiences of injustice when other agents rely on information from these systems to guide their actions, often within partially or fully AI-driven, institutionalized decision structures (Véliz 2021). This provokes resistance among the misrecognized and negatively affected, expanding the moral vocabulary of contemporary protest movements. In this way, AI operates on two levels simultaneously: it reshapes the conditions under which recognition relations and struggles unfold, while becoming a contested object around which new claims for justice, dignity, and agency crystallize.

Theorizing AI systems as emerging objects of conflict suggests that contemporary recognition struggles center on questions of technological governance: how these systems are coded, deployed, owned, and regulated. The stakes involve preventing privacy violations, protecting marginalized communities from exclusion, and constraining malicious applications. Crucially, these struggles target the underlying power structures—commercial ownership as

---

[7] Waymo, owned by Alphabet (Google's parent company), has also been criticized as a symbol of AI-driven job displacement, further highlighting the complex intersections between technological development, civil liberties, and rapid socio-economic change.

well as business models, labor processes, and decision-making frameworks—that determine AI's impact on lived social experience, seeking to establish chains of ethical responsibility. This perspective shifts focus away from popular debates about AI's potential autonomy or moral agency and toward another persistent concern: how these technologies are reconfiguring human-to-human and human-to-institution recognition relations through new forms of technical mediation. Regardless of the form these systems take, the critical issue remains their role in transforming the conditions under which recognition is practiced and negotiated across social domains. Scholars of mediated recognition have begun to trace these developments (Campanella 2024; Jansson 2015; Driessens and Nærland 2022), examining how AI's growing capacities for perception, classification, language generation, and prediction reshape the ways recognition is expressed, received, and valued—well beyond the scope of traditional face-to-face interaction.

The following section examines concrete cases that demonstrate these dynamics—specifically, how recognition practices mediated through AI-driven social media platforms are reshaping protest movements and their struggles for recognition in an increasingly digitalized society.

2.2 AI Systems as Means of Struggle

Contemporary protest movements increasingly rely on AI-powered social media platforms—from hashtag amplification to content virality to targeted event advertising—as means to advance claims, mobilize supporters, and bring about change. Yet these same platforms are hosted by corporations like Meta, X, and ByteDance, which operate advanced AI systems that monitor user behavior, generate predictive knowledge, and monetize these data for targeted advertising, including political campaign advertisements. This business model, often termed "surveillance capitalism" (Zuboff 2019) or "data colonialism" (Couldry and Mejias 2019), harnesses massive machine learning capabilities to extract undisclosed information from user posts and detect hidden patterns across vast datasets. These dynamics raise crucial issues for contemporary protest movements: can the same AI-driven platforms that enable activists to achieve social visibility, mobilize support, and challenge existing power relations simultaneously constrain and undermine recognition struggles? The answer reveals AI's dual function as both enabling and constraining means within digitalized recognition politics.

The enabling dimension becomes evident when AI systems amplify protest visibility through sophisticated content curation and distribution.

Machine-learning algorithms analyze social-media trends, enabling users to tailor their messages for maximum impact (Polyakova and Meserole 2019), while hashtags serve as discursive tools that organize participants, frame issues, and increase visibility through what Bennett and Segerberg (2013) call connective action. This was powerfully demonstrated during the 2020 Black Lives Matter protests following George Floyd's death, when Darnella Frazier's footage of police officer Derek Chauvin went viral on Facebook within 48 hours (Huq 2021). The hashtags #JusticeforGeorgeFloyd and #BlackLivesMatter were algorithmically amplified across platforms, with recommendation algorithms pushing these narratives across networks, transforming public discourse into a global movement. The algorithmic logic of engagement worked in activists' favor by amplifying trending content, democratizing access to public attention previously managed by traditional media. Over 44 million #BlackLivesMatter tweets from nearly 10 million distinct users illustrated how AI systems function as amplifying tools in digitalized struggles for recognition (Khalil 2020), empowering marginalized communities[8] to directly communicate their struggles to broader publics.

However, this technological empowerment comes with significant trade-offs. While AI systems enable protest visibility, they simultaneously impose constraints that may distort and fragment collective struggles for recognition. Recent scholarship illuminates how machine learning algorithms reshape recognition practices on digital platforms. For instance, Bruno Campanella (2022) introduces "datafied recognition" to assess how digital infrastructures reshape media visibility, while Jacobsen (2021) examines "regimes of recognition" through X's deep saliency algorithm—a machine learning system that automatically crops images by deeming some aspects of an image more salient or more irrelevant to the user.[9] These algorithmic regimes govern what Jacobsen terms the conditions of possibility determining who and what is seen, attended to, and made to matter through recommender systems that personalize content and interface experiences. As Jacobsen argues, platforms "algorithmically shape and organize people's parameters of attention, and in turn, create subjects that are productive, predictable, and perpetually compliant" (Jacobsen 2021, 13). This dynamic becomes evident when activist

[8] Research (Craig et al. 2020; McInroy and Craig 2017) also indicates that social media platforms may empower LGBTQ+ youth to explore their identities and build supportive networks within affirming online environments.

[9] X's deep saliency algorithm is a machine learning system that determines content visibility and ranking in users' feeds by analyzing engagement patterns, user behavior, and content characteristics to predict and prioritize what will capture user attention. "Saliency" refers to the algorithm's assessment of content's attention-capturing potential and relevance to individual users, determining what receives prominence in their feeds.

content is shadow-banned or when algorithms prioritize sensational over substantive political discourse, effectively filtering out voices that challenge dominant narratives.

Political economist Will Davies (2021) identifies a critical paradox within engagement-driven social media algorithms, beyond online content curation. While these systems enable unprecedented mobilization capacities, they simultaneously risk fragmenting the foundations of collective political action. Davies argues that algorithmic engagement structures incentivize activists to adopt platform commodification logics, transforming substantive political struggles into gamified contests for online visibility. Algorithms distribute recognition through competitive metrics—likes, shares, followers—fostering winner-take-all dynamics where "a few nodes receive an abundance of connectivity and engagement while the vast majority receives very little" (2021, 96 f). The result is what Davies describes as distorted struggles over mere public reputation, where individual esteem accumulation displaces collective solidarity and shared claims for equal respect-recognition (Davies 2021, 92 ff.). Rather than challenging existing hierarchies of social worth, algorithmic engagement systems may reproduce and intensify them, turning the platform's promise of democratic participation into a mechanism that fragments collective action into individualized competitions for visibility, incentivizing what Naaman et al. (2010) term 'me-formers'; users focused on broadcasting mere personal status rather than engaging in political messaging, potentially neutralizing the power of progressive movements.

Thus, beyond issues of technological governance, theorizing AI systems as means of recognition struggles clarifies how these technologies shape activist practices. Algorithmically curated social media platforms function as communicative tools for articulating recognition claims—tools that technologically mediate recognition exchanges while creating new experiential environments for such interactions—and processes that resemble traditional practices of mutual confirmation while altering how identity, respect, and social justice demands are articulated, received, and collectively pursued. Yet, this analysis also reveals that AI-mediated social recognition constitutes a double-edged sword for contemporary protest movements. While algorithmic systems can enable democratic access to public attention, they can likewise function as constraining means that embed competitive logics, potentially undermining collective solidarity.

These systems create contradictory effects. They may amplify individual testimonies into mass movements but simultaneously fragment these movements into individualized struggles for visibility. For activists navigating these technological means, the challenge is not simply choosing between engagement and withdrawal, as retreating from traditional platforms or building counter-platforms often appear unrealistic in the short term given existing infrastructures' scale and reach. Instead, activists must grapple with the reality that these AI-powered technologies operate according to algorithmic logics that prioritize engagement over solidarity, metrics over meaning, and individual reputation over joint action. This requires developing new forms of political practice that harness these technological means and their amplifying potential while resisting their fragmenting effects—grasping that recognition struggles require preserving collective agency within algorithmically curated social media platforms designed to individualize and commodify content.

3 Algorithmic Exploitation of Recognition-Needs

3.1 Recognition Vulnerabilities in Social Media

Thus far we have argued that AI-powered social media platforms resemble traditional recognition practices by operating across two fundamental dimensions. At the cognitive level, AI systems function as sophisticated pattern identification technologies—exemplified by facial recognition systems that detect and classify human features. At the practical level, these systems technologically mediate the processes through which individuals value, esteem, and respect one another, serving as digital tools for recognition exchanges. Furthermore, we have seen that AI-powered social media platforms may give rise to and shape recognition struggles in numerous ways: (1) AI itself becomes an object of struggle due to new forms of misrecognition; (2a) AI systems provide means for rapid mobilization, offering marginalized voices opportunities for amplified visibility; and (2b) these same technological means risk converting recognition struggles into pathological competitions for public reputation and individualized esteem.[10] This analysis uncovers both the enabling and the constraining tendencies inherent in AI-mediated recognition struggles. The future of these struggles depends on whether activists gain awareness of these dual tendencies and can sustain collective solidarity necessary for progressive societal change while navigating algorithmically curated

---

[10] AI systems functioning as tools within conflicts can constitute objects of struggle. This is evident when state actors use AI systems to surveil and suppress dissenting voices online, prompting civil society demands for democratic oversight and regulatory constraints (see Shahbaz 2021).

social media platforms that prioritize individual engagement over collective action.

Despite extensive scholarship on movement protest and digital activism (e.g., Tilly and Tarrow 2015; Jackson et al. 2020; Milan 2013; Celikates 2015; Delmas 2018; Moody-Adams 2022; Haslanger 2012) and emerging work in the philosophy of protest (Medina 2023; Delmas and Brownlee 2021), theorists have largely overlooked how AI systems exploit fundamental human needs for affirmation to drive platform engagement and maximize profit. Building on Davies' analysis of distorted struggles over reputation, this section examines how social media platforms and their underlying AI systems affect struggles for recognition through the manipulation of recognition needs. While much criticism focuses on how these AI-powered platforms exploit cognitive biases to capture attention, this analysis reveals a more insidious form of exploitation. That is treating vital recognition needs—our needs for affirmation and social validation—as psychological vulnerabilities to be monetized. This exploitation operates transparently, as Sean Parker, former Facebook president, acknowledged when describing the deliberate design principles behind these platforms:

> The thought process [...] was all about: 'How do we consume as much of your time and conscious attention as possible?' [...] 'we needed to give you a little dopamine hit every once in a while, [...] It's a social-validation feedback loop [...] you're exploiting a vulnerability in human psychology [...] we did it anyway." (Parker, quoted in Fried 2017)

There is mounting evidence that the quest for affirmation and reward constitutes the principal energizing source for online media engagement (Eyal 2019)—a psychological dynamic that has long been recognized and exploited by user interface designers and online marketers engaging in mood prediction for targeted advertising (Tiffany 2019). In fact, AI-enabled exploitation of recognition needs operates across many dimensions, as Honneth's normative reconstruction approach helps reveal (Honneth 2014, 10). AI-enabled platforms target individual cravings for reward and reputation (esteem-recognition) through features like following, liking, or group confirmation, while simultaneously exploiting desires for personal attachment and mutual affection (love-recognition)—as evidenced by the commercial success of dating platforms like Tinder. Perhaps most significantly, they manipulate fundamental needs for bonding and belonging as equal community members (respect-recognition) through participation in like-minded groups on social media. Facebook's introduction of the "like" feature exemplifies this approach: engagement soared as users received affirmation boosts while the company collected valuable preference data for algorithmic predictions (Lewis 2017).

Yet, recognition theorists have largely neglected the normative and conceptual issues raised by AI-enabled exploitation of recognition-needs for profit purposes, with a few exceptions in media studies (e.g., Jansson 2015; Magalhães and Yu 2022), despite growing concern about how AI-driven platforms prioritize affective interaction, amplifying emotional mobilization and moral outrage rather than reasoned argument (Gerbaudo 2022, 122).

How should we conceptualize this AI-enabled exploitation of recognition needs? While there is much appeal in Marx' notion of exploitation (see Venkatesh 2021)—extracting surplus value from human labor—the concept has been revived recently with a broader and more normative understanding rooted in Kantian ideals—which prohibit treating people as mere means to financial ends, neglecting their intrinsic dignity (Mills 2018, 5; Zwolinski et al. 2022). Following this notion, we understand AI-enabled exploitation as a form of instrumental rationality that treats people's vital needs for affirmation as mere means to their operators' financial ends. This exploitation operates through specific mechanisms: AI systems monetize recognition needs through unprecedented technological means. Digital platforms reveal distinct business models that systematically convert human desires for social validation into profit-generating engagement patterns. Unlike the old business model described in Brennan and Pettit's *Economy of Esteem* (2004), where recognition rewards directly motivated customer behavior, AI-enabled platforms operate through a more sophisticated model. Instead of explicit transactions, these systems promise recognition-need satisfaction in exchange for personal data, using AI's unique predictive capacities to determine precisely when rewards should be supplied to effect user behavior. As design theorist Eyal (2014) demonstrates, these rewards and related affirmation experiences drive user engagement patterns, transforming users into what Shoshana Zuboff terms the raw material of prediction products—user recognition needs become inputs for algorithmic systems designed to modify future behavior rather than to simply satisfy present desires. The centrality of recognition needs to digital platform operations has not gone unnoticed. As Zuboff observes in *The Age of Surveillance Capitalism* (2019):

> (…) This process can be accomplished successfully only in the presence of our unrelenting hunger for recognition, appreciation, and most of all, support. (Zuboff 2019, 255)

Therefore, what distinguishes AI-enabled recognition exploitation from traditional forms is its unprecedented scale, precision, and systematicity. It has long been known that what draws actors into markets is not only maximizing returns (Friedman 1970), but also enjoying the merit and reward given by others for individual contributions to cooperative aims.[11] Unlike human actors who may exploit recognition needs opportunistically (whether through complimenting others to nudge them to do X, or through promises of rewarding feelings if they purchase Y), AI systems are engineered to detect, predict, and manipulate recognition vulnerabilities continuously and at industrial scale. These algorithmic recognition processes strategically time, target, and curate recognition exchanges between humans to maximize engagement and data extraction rather than to facilitate genuine social connection. Unlike traditional mediated recognition, where technology simply facilitates human-to-human affirmation,[12] algorithmic recognition involves AI systems manipulating when, how, and to whom recognition is distributed through personalized, algorithmically recommended content and persuasive interface designs based on predictive models of user vulnerability. This reveals the core mechanism of AI-enabled recognition exploitation: users experience what feels like authentic social validation while simultaneously providing behavioral data that platforms use to optimize future manipulation of their recognition needs. The apparent reciprocity—for example, likes for engagement, community for data—masks the fundamental power asymmetry defining platform-user relationships.

Dating apps monetize love-recognition vulnerabilities through premium features like super-likes or roses. While dating platforms like Tinder extend beyond traditional social media, they operate through similar algorithmic recognition mechanisms that exploit fundamental human needs for affirmation and connection. When users experience rejections, AI systems strategically promote premium features during vulnerable moments, offering purchased validation that promises temporary recognition (Binder et al. 2024). Similarly, TikTok exploits esteem and respect-recognition vulnerabilities through its

[11] The link between recognition and markets (Celentano and Caranti 2020) was best addressed in the exchange between Fraser and Honneth, *Redistribution or Recognition?* (2003). First, the underlying idea is that what draws actors into markets is not only maximizing returns (Friedman 1970), but also enjoying the merit given by others for individual contributions to shared and cooperative aims. Summarizing Honneth's view, markets are normative orders that embody merit-based esteem—what and whom we value as a society—and that related norms affect how we organize the allocation of goods, the division of labor, and the distribution of the burdens and benefits of cooperation (Honneth 2014, 183 f).

[12] There remains a fundamental ontological question about whether these technologically mediated forms of feedback can be considered recognition at all, given what philosopher John Durham Peters calls the irreducibility of touch and time: "all things being equal, people who care for each other will seek each other's presence" (Peters 1999, 270 f.). For broader discussion of AI systems and recognition's relational dimensions, see chapters two and three in this volume.

'For You Page.' Their algorithmic systems track user interactions such as likes, comments, or watch time, while also using Natural Language Processing to analyze text and Computer Vision to detect emotions in videos. This sentiment analysis feeds into TikTok's recommender systems, promoting emotionally resonant content and encouraging creators to shape their material for validation through likes and shares. In both cases, the platforms monetize human vulnerabilities by offering validation that promises both individual recognition and community belonging but delivers only temporary and algorithmically manipulated recognition designed to extract engagement data, perpetuating cycles of exploitation (Bhandari and Bimo 2022).

Recent developments suggest these algorithmic mechanisms may be evolving toward more sophisticated forms of manipulation. Emerging phenomena like AI companions and AI avatars that conduct dates on behalf of users (e.g., Baradari et al. 2025) point toward what we term 'synthetic recognition'—where AI systems simulate being recognizing subjects themselves, providing validation responses rather than merely mediating human exchanges. Such developments indicate that platforms may move beyond algorithmic curation toward AI systems that manufacture pseudo-social responses designed to exploit recognition needs for engagement rather than express genuine appreciation.

This raises a crucial issue: why is AI-enabled exploitation of recognition needs objectionable? Benefiting from vulnerability is not inherently wrong—chess players exploit opponents' weaknesses without ethical concern. Moreover, some online experiences may genuinely satisfy recognition desires. What distinguishes legitimate and authentic recognition from exploitative algorithmic recognition, and when does such exploitation become problematic? To address these questions, the next section examines how AI-driven social media platforms exploit psychological vulnerabilities in ways that erode personal autonomy through four distinct forms of unjustified misrecognition that render such exploitation ethically objectionable.

3.2 The Moral Failures of Algorithmic Recognition

Why is AI-enabled exploitation of recognition objectionable? The answer lies in understanding how these systems violate the normative foundations of authentic recognition practices.[13] While algorithmic recognition may appear

[13] Authentic recognition practices involve reciprocal relationships between subjects who mutually shape themselves in light of each other's claims (Honneth 2018, 187ff). This contrasts with algorithmic recognition, which treats users as passive objects of manipulation rather than active participants in determining appropriate forms of acknowledgment.

to satisfy users' needs and even nudge individuals toward seemingly beneficial purchases, closer examination reveals at least four distinct forms of misrecognition that render such exploitation morally problematic and theoretically significant for contemporary recognition struggles. These forms of misrecognition both draw upon and contribute to the current exchange between recognition theory and critical social epistemology (Giladi 2023).

The first form of misrecognition manifests as *structural disrespect for informational self-determination*: a failure to recognize users as autonomous agents capable of making informed decisions about their personal data and digital identity. AI-driven platforms misrecognize users by treating them as passive data sources rather than as subjects deserving respect for their informational autonomy. Users become passive participants in extensive data-gathering processes that operate without their knowledge or meaningful consent. Most algorithmic data collection—including users' sex, age, nationality, political affiliations, or preferences—occurs through sophisticated inference mechanisms that extract information from seemingly innocuous activities such as online search behavior, click patterns, and social interactions (Hinds and Joinson 2018; Metzler and Garcia 2024). This creates profound harm when users cannot exercise meaningful control over what personal information is extracted, particularly when they explicitly do not want to share such data. The violation extends beyond simple privacy concerns to encompass a fundamental assault on users' autonomy over their digital identity and self-presentation. Users become unknowing contributors to their own surveillance, trapped within systems that systematically undermine their capacity for self-determination in the digital sphere.

The second form of misrecognition operates as *disregard for epistemic agency*, constituting what Miranda Fricker (2007) might recognize as hermeneutical injustice applied to the digital realm (Medina 2018). This manifests through systematic failure to recognize users as legitimate knowers deserving access to interpretive resources needed to understand how their datafied existence is constructed. While platform companies deploy capable AI systems to construct detailed "statistical online doubles" (Origgi and Ciranna 2017, 307) or "data doubles" (Hepp 2020, 158–160), users are largely kept in the dark about these processes. Platforms operate as 'black boxes' that render users' experiences of algorithmic manipulation incomprehensible, denying them the conceptual tools necessary to make sense of how their recognition needs are being exploited. This epistemic asymmetry creates a power imbalance wherein corporate entities possess knowledge about users' behavioral patterns, preferences, and potential psychological vulnerabilities, while users themselves lack access to this knowledge about their own digital representations.[14]

Moreover, the testimonial injustice that users face in surveillance capitalism operates through the systematic discrediting of their testimony about themselves in favor of quantified representations of their online behavior—a phenomenon that extends Fricker's (2007) analysis of testimonial injustice into the digital domain. The algorithmic interpretation of user data becomes more authoritative than users' own self-understanding, effectively undermining their credibility as witnesses to their own experiences and intentions. Both forms of epistemic misrecognition—hermeneutical and testimonial—harm users' capacities for self-knowledge and self-understanding, making them less credible as witnesses of their own lives and challenging the conditions necessary for autonomous agency. This epistemic disempowerment renders users vulnerable to exploitation by undermining their ability to understand and resist the systems that seek to manipulate them (Roessler 2020, 205 f).

The third form of misrecognition involves the *violation of democratic participation in recognition practices themselves*. We can understand how AI-enabled exploitation undermines this principle by drawing on Honneth's analysis in *Recognition and Ambivalence* (Ikäheimo et al. 2021), where he argues in discussion with Judith Butler that mutual recognition in a normative sense requires democratic participation. Honneth contends that those subjected to recognition norms and practices must be able to engage as co-authors of these norms, with both recognizing and recognized subjects having meaningful input into what appropriate respect, esteem, or love requires—particularly for those being recognized. This democratic element ensures that recognition practices remain responsive to the actual needs and perspectives of those they purport to serve (Honneth 2021, 55–60).

AI-enabled recognition exploitation profoundly violates this democratic principle by establishing recognition norms and practices through algorithmic systems that operate entirely without user input or democratic deliberation. Platform companies unilaterally determine what constitutes valid recognition, how recognition should be distributed, and what behaviors merit social validation, while users remain excluded from any meaningful participation in shaping these social processes. Users cannot participate as co-authors of the recognition practices that increasingly govern their social lives; instead, they find themselves subjected to recognition regimes designed primarily to

[14] Within the EU, the 2018 General Data Protection Regulation (GDPR) has granted individuals the rights to access the data that platforms collect about them (GDPR Article 15) and to data portability (GDPR Article 20), enabling so-called data donation research approaches where citizens voluntarily donate their platform data for research purposes (e.g., Kormelink et al. 2025).

serve corporate rather than human interests. This represents a misrecognition of users as democratic agents capable of participating in the construction of their own social worlds.[15]

The fourth form of misrecognition involves the *objectification of recognition subjects themselves*. This represents the most insidious failure of algorithmic recognition systems: the systematic misrecognition of users as mere objects for manipulation rather than as subjects deserving genuine acknowledgment and respect. This objectification operates through a peculiar split between the cognitive and practical dimensions of recognition that reveals the pathological nature of AI-mediated validation. While these systems may demonstrate considerable sophistication in accurately detecting users' vital needs for belonging, validation, and esteem—thus appearing to satisfy the cognitive dimension of recognition through what we might call algorithmic recognition—they systematically fail to honor the practical dimension of recognition that involves genuinely valuing, appreciating, and respecting others as autonomous subjects deserving of care and consideration.

This failure manifests most clearly in how platforms treat recognition needs as mere variables to be optimized for engagement and profit extraction rather than as legitimate expressions of human dignity deserving appropriate normative response. Drawing on Honneth's analysis of reification, we can understand this as a particularly pathological form of social interaction where users are not genuinely 'seen' or acknowledged as full persons with intrinsic worth but rather are processed as collections of psychological triggers and behavioral response patterns to be manipulated for commercial ends. In *Reification* (2008), Honneth argues that objectification occurs when we lose sight of the intersubjective nature of human existence and begin to treat others as objects to be manipulated rather than as beings with intrinsic worth.[16] The recognition provided by these systems becomes purely transactional and instrumental: users receive carefully calibrated algorithmic validation designed to maintain engagement while their fundamental needs for authentic social connection are systematically subordinated to corporate interests in data extraction and profit maximization.

---

[15] The three forms of misrecognition outlined above can also be understood in terms of an undermining of the conditions for the three aspects of Honneth's (2014) Hegelian conception of freedom (i.e., individual autonomy)—negative, reflexive, and social freedom, respectively (see Magalhães and Yu 2022, for an analysis of social media platforms' role in this regard).

[16] Drawing on Lukács' notion of reification, Honneth develops objectification as a form of failed recognition in which humans are treated as mere things rather than as subjects deserving of acknowledgment and respect. In the context of recognition, objectification represents a pathological form of social interaction where the recognizing party fails to acknowledge the recognized party's subjectivity, autonomy, and inherent dignity.

This represents what Honneth describes as objectification in its most sophisticated contemporary form. The algorithmic exploitation of recognition needs thus constitutes a form of reified social interaction that hollows out the intersubjective foundations upon which meaningful recognition practices depend. Users find themselves trapped within systems that algorithmically curate or even simulate the forms of social validation they genuinely need while systematically undermining the conditions that would make such recognition authentic and mutually enriching.

The moral significance of these failures extends beyond individual harm to encompass broader implications for the social fabric itself. When basic human needs for esteem, love, and respect become raw material for algorithmic optimization, the very possibility of authentic recognition practices becomes compromised. Users learn to seek validation through channels designed to exploit rather than fulfill their social needs, potentially developing pathological attachments to forms of pseudo-recognition that ultimately leave them more isolated and vulnerable than before—a dynamic that Magalhães and Yu (2022) term institutionalized social unfreedom. This degradation of recognition practices represents not merely a technological problem requiring better regulation, but a major challenge to the intersubjective foundations of human flourishing in digital societies.

One might object that all forms of recognition are potentially problematic and dominating, rendering the specific critique of AI-enabled exploitation less compelling. This objection draws on what Lepold (2019) identifies as the tension between positive and negative concepts of recognition. While Honneth employs a positive concept of recognition, viewing acts of recognition as crucial to enabling individual autonomy and social inclusion, negative accounts regard recognition as promoting conformism and dependency (Markell 2003; Bertram and Celikates 2015; Ikäheimo et al. 2021, 3 f).

Central to this disagreement is whether we should understand acts of recognition as either ascribing or responding. The ascriptive model places the recognizer in a privileged social position, controlling the recipient's development of personal identity by ascribing positive and negative characteristics, while the responsive model privileges the recognized agent, analyzing recognition as perceiving and responding to objective characteristics in the recipient that merit acknowledgment (Zurn 2015, 52). Negative accounts typically endorse the ascriptive model and reduce recognition acts to forms of labeling others with dominant norms.

However, Honneth's positive account offers a more differentiated middle position particularly relevant for understanding AI-enabled exploitation. Honneth assigns both parties—recognizer and recognized—crucial roles in successful recognition acts. Recognition acts must correspond to actual characteristics worthy of acknowledgment, while recipients must realize and internalize the evaluative attitude to constitute their social characteristics. Therefore, both recognizer and recognized play essential roles in the dynamic tension between other-regarding attitudes and the development of positive social identity (Zurn 2015, 54).

Applying this framework to AI systems reveals why algorithmic recognition exploitation is particularly pernicious. Unlike authentic recognition practices that involve genuine responsiveness between recognizing and recognized parties, algorithmic recognition systems operate through pure ascription without any meaningful responsiveness to users' actual characteristics or legitimate needs for acknowledgment. AI systems ascribe recognition rewards based not on users' genuine deeds or inherent worth, but on their value as sources of behavioral data, engagement metrics, and potential profitability. Users cannot participate as genuine partners in recognition exchanges, nor can they influence how their characteristics are perceived and evaluated by algorithmic systems. The result is an asymmetrical relationship that violates the basic requirements of authentic recognition while maintaining its external appearance. AI-enabled exploitation is objectionable not simply because recognition can be problematic in general, but because it systematically eliminates the responsive, democratic elements that distinguish authentic recognition from mere domination.

Unlike chess players who mutually understand their competitive framework, AI systems exploit recognition vulnerabilities through deception and asymmetrical power relations that users cannot meaningfully contest. While some online experiences may indeed satisfy recognition desires, the four forms of misrecognition identified here reveal that such satisfaction comes at the cost of users' informational autonomy, epistemic agency, democratic participation, and basic dignity as subjects rather than objects. What distinguishes authentic from algorithmic recognition is whether the recognition practices preserve the responsive, democratic, and genuinely intersubjective foundations that make recognition a vehicle for human flourishing rather than exploitation.

3.3 From Exploitation to Resistance: Emerging Recognition Politics

The four forms of misrecognition that underpin AI-enabled exploitation reveal how recognition itself can degrade into a highly sophisticated transactional relationship, where one side benefits by exploiting another's vulnerabilities. Yet awareness of this emerging form of exploitation can ignite new forms of resistance, challenging the foundations of how recognition operates in digital environments. In fact, we believe that there are already at least three emerging patterns of struggle illustrating these new dynamics and indicating likely future developments in digitalized recognition politics.

The first pattern is *resistance to social media addiction* among youth, which reveals how awareness of recognition exploitation generates collective action against addictive cycles created by metrics-driven validation (Haidt 2024). With teenagers spending 7–9 hours daily on screens (Rideout et al. 2022) and checking phones over 150 times per day, seeking both infotainment and peer group validation (Asurion 2019), there is growing awareness of how platforms exploit young people's needs and vulnerabilities through quantified metrics—likes, shares, or followers. This has sparked initiatives such as digital detoxes (Hunt et al. 2018; Syvertsen 2020) and the Time Well Spent movement (Center for Humane Technology, n.d.), and a range of other disconnection practices (Albris et al. 2024) as well as platform migrations such as the exodus from WhatsApp to Signal (Reuters 2021). Further, regulatory responses like Australia's proposed Social Media Minimum Age Bill in 2024 restricting access for children under 16 reflect growing institutional awareness that protecting youth from recognition exploitation requires systemic intervention. These initiatives show resistance to AI-enabled validation designed to create dependency rather than real-world engagement, and demand respect for epistemic agency, challenging the subordination of human needs for social affirmation to algorithmic performance metrics.

The second pattern emerges in *gig worker resistance to algorithmic rating systems*, revealing how recognition exploitation extends beyond social media platforms to workplace validation techniques. When Uber and Lyft drivers organize against rating systems that algorithmically process user feedback to reduce them to behavioral scores (Rosenblat 2018), or when delivery workers resist surveillance metrics (Cant 2019), they fight for recognition as social agents rather than objects of algorithmic evaluation. These struggles reveal how the objectification of recognition subjects operates through the broader infrastructure of the emerging AI-enabled digital economy. Workers demand transparency about performance evaluation, due process when ratings lead to deactivation, and meaningful participation in establishing appropriate recognition of their labor. This new workplace struggle goes beyond traditional wage demands to encompass human dignity and democratic oversight over recognition practices in algorithmically structured work environments. As AI systems increasingly mediate employment relationships, we can anticipate expansion of such recognition struggles where workers demand authentic acknowledgment rather than reduction to manipulable data points.

The third pattern reveals *algorithmic amplification and weaponization of political grievances*, showing how movements like MAGA exploit unaddressed recognition needs for political ends. The "Make America Great Again" slogan operates as a recognition appeal, promising restored status to those feeling marginalized (e.g., Koenig and Mendelberg 2025; Hochschild 2016). Rather than addressing structural sources of marginalization, documented cases reveal how political campaigns systematically exploit these recognition needs through AI-powered algorithmic targeting systems. The Cambridge Analytica scandal exemplifies this exploitation most prominently, showing how campaigns used Facebook's algorithmic advertising platform combined with psychological profiling to micro-target users experiencing economic anxiety, social isolation, or status threat (Rosenberg et al. 2018). These AI systems analyzed user data to identify recognition vulnerabilities, then deployed targeted ads that validated users' grievances while channeling recognition needs into zero-sum competition against vulnerable populations and sparking polarization (Benkler et al. 2018). Overall, we argue that online MAGA communities create recognition reward systems where likes and shares become validation currency, while extreme anti-immigrant posts receive algorithmic amplification (Tucker et al. 2018).

These examples reveal several crucial theoretical insights about recognition in the age of AI. First, they demonstrate that AI-enabled exploitation creates a new category of recognition struggle focused on the infrastructure and governance of recognition itself rather than its content or distribution. Second, they show how the four forms of misrecognition identified earlier operate across multiple social domains, suggesting that recognition exploitation represents a systemic rather than an isolated phenomenon. Third, they illustrate how both the awareness of and the suffering from such exploitation (e.g., addiction, or being marginalized) can generate both progressive and reactionary political mobilization, highlighting the contested nature of recognition struggles. Also, they reveal that individuals and communities must resist systems designed to weaponize their fundamental need for recognition. This represents a meta-level of recognition struggle: conflicts over who controls the infrastructure of recognition, how recognition needs are identified and used, and whether recognition practices are structured democratically and with respect for individuals' epistemic agency.

The theoretical contribution of this analysis lies in demonstrating how AI fundamentally transforms recognition politics by creating novel forms of domination that operate through the corruption of recognition practices themselves. The four specific forms of misrecognition identified as underlying algorithmic exploitation reveal how technology does more than provide new tools for existing struggles: it generates entirely new dynamics where the infrastructure of recognition becomes a site of contest. The stakes extend beyond individual validation to encompass the possibility of democratic solidarity in digital societies. When recognition needs become raw material for algorithmic optimization, the social conditions necessary for collective action may be degraded into competitive performance for algorithmic validation. Addressing this challenge requires better AI regulation, as well as sustained efforts to create technological infrastructures that serve authentic human recognition needs rather than exploiting them for corporate profit.

Looking forward, these developments suggest the emergence of new struggles demanding what we might call "algorithmic recognition rights"—legal and political frameworks ensuring that AI-assisted systems respect human dignity, enable democratic participation in recognition practices, and preserve authentic intersubjective acknowledgment (see also Coeckelbergh 2022; Dubber et al. 2020). Additionally, these developments may give rise to conflicts over proper respect for epistemic agency and personal freedom; that is, struggles to control the level of disclosure of personal data and secure the right to sanctuary in the age of surveillance capitalism (Zuboff 2019, 475 ff).

4 Conclusion: Reclaiming Recognition from Algorithmic Exploitation

This chapter has demonstrated that AI systems transform recognition politics by operating across cognitive and practical dimensions of recognition practices. At the cognitive level, these systems function as pattern identification technologies—exemplified by facial recognition that classifies human features. At the practical level, they mediate social processes through which individuals value, esteem, and respect one another, creating new digital tools and spaces for recognition exchanges. Yet this reveals a crucial distinction: while traditional recognition involves authentic human-to-human acknowledgment, AI systems generate algorithmic recognition—algorithmically curated responses that mediate human interactions while serving extractive purposes rather than facilitating genuine validation.

Our analysis has revealed that AI-powered social media platforms shape recognition struggles in four ways. First, AI becomes an object of struggle due to new forms of misrecognition, as communities organize against biased algorithms that systematically misclassify and exclude marginalized groups. Second, AI systems function as means of struggle, providing tools for mobilization and offering marginalized voices opportunities for amplified visibility, as demonstrated by #BlackLivesMatter. Third, these same technological means may convert authentic recognition struggles into pathological competitions for individualized esteem, undermining collective solidarity necessary for transformative change. Fourth, we have identified emerging patterns of both resistance to and weaponization of recognition exploitation, revealing the contested nature of recognition struggles in digital environments.

This chapter has sought to address a critical gap in understanding how AI-enabled digital environments affect struggles for recognition and give rise to new forms of struggle. While much attention has focused on how AI systems misclassify or exclude through algorithmic misrecognition, our investigation demonstrates that these technologies pose a more fundamental threat through the algorithmic exploitation of recognition needs themselves. Unlike simple failures of identification, this exploitation operates by converting vital human needs for affirmation into raw material for profit extraction through algorithmic recognition mechanisms. Most paradoxically, activists find themselves relying on tools designed to extract profit from human needs for affirmation, creating tension between social justice organizing and the extractive logic of AI-driven digital platforms.

This algorithmic exploitation operates through four distinct forms of misrecognition we have identified—violations of informational self-determination, disregard for epistemic agency, denial of democratic participation, and objectification of subjects—which together represent novel challenges to social freedom in the digital age. When AI systems exploit recognition needs for corporate profit, they undermine the intersubjective foundations necessary for individual autonomy and collective democratic action. Recognition becomes raw material for algorithmic optimization, threatening to degrade social bonds essential for solidarity and political action.

This transformation poses fundamental questions about social freedom in AI-driven societies. If human needs for esteem, respect, and belonging become systematically manipulated by algorithmic systems, the possibility of authentic recognition faces unprecedented challenges. These emerging struggles over recognition suggest that preserving social freedom requires protecting individuals from algorithmic harm and creating technological infrastructures that serve authentic human recognition—challenges that will intensify as AI becomes even more immersive. As algorithmic recognition evolves toward synthetic recognition, the implications extend beyond current platforms to encompass pervasive AI technologies embedded in daily life. Questions emerge about how recognition struggles might unfold in increasingly immersive AI-enhanced environments, including metaverse spaces where the boundaries between human and AI-generated interactions may become indistinguishable, and about how emerging technologies like relational AI and ambient computing might fundamentally reshape the conditions of daily recognition practices. As AI systems become more sophisticated at simulating recognitive responses, the boundaries between authentic and synthetic recognition may blur beyond current theoretical frameworks.

Acknowledgements The authors thank Christopher F. Zurn (University of Massachusetts Boston), Anat Biletzki (Quinnipiac University), Emmanuel Renault (Université Paris Nanterre), Rosalie Waelen (University of Bonn), Jesse Tomalty, Franz Knappik, Hallvard Moe (University of Bergen), and Lauren McGillicuddy for their insightful comments and critical feedback that substantially improved earlier drafts of this chapter.

References

ACLU. 2020. *ACLU Statement on Amazon Face Recognition Moratorium*. Press release, June 9, 2020. https://www.aclu.org/press-releases/aclu-statement-amazon-face-recognition-moratorium.

Albris, Kristoffer, Karin Fast, Faltin Karlsen, Anne Kaun, Stine Lomborg, and Trine Syvertsen, eds. 2024. *The Digital Backlash and the Paradoxes of Disconnection*. Nordicom: University of Gothenburg. https://doi.org/10.48335/9789188855961.

Algorithmic Justice League (AJL). 2025. *Unmasking AI Harms and Biases*. Accessed 21 Jun 2025. https://www.ajl.org.

Amazon Web Services (AWS). 2025a. *Amazon Rekognition: Image Recognition Software, ML Image & Video Analysis*. Accessed 21 Jun 2025. https://aws.amazon.com/rekognition/.

Amazon Web Services (AWS). 2025b. *Face Collections*. Amazon Rekognition Developer Guide. Accessed 21 Jun 2025. https://docs.aws.amazon.com/rekognition/latest/dg/collections.html.

Asurion. 2019. *Americans Check Their Phones 96 Times a Day*. Press release, November 21, 2019. https://www.asurion.com/about/press-releases/americans-check-their-phones-96-times-a-day/.

Australian Government. 2024. *Social Media Minimum Age Bill 2024*. Commonwealth of Australia.

Baradari, Dünya, Tejaswi Polimetla, and Pattie Maes. 2025. Data-Driven AI Avatars for Valuation in Dating Scenarios. In *Proceedings of the Extended Abstracts of the CHI Conference on Human Factors in Computing Systems*, 1–13.

Bartoletti, Ivana. 2020. *An Artificial Revolution: On Power, Politics and AI*. London: Indigo Press.

Benkler, Yochai, Robert Faris, and Hal Roberts. 2018. *Network Propaganda: Manipulation, Disinformation, and Radicalization in American Politics*. New York: Oxford University Press.

Bennett, W. Lance, and Alexandra Segerberg. 2013. *The Logic of Connective Action: Digital Media and the Personalization of Contentious Politics*. Cambridge: Cambridge University Press. https://doi.org/10.1017/CBO9781139198752.

Bertram, Georg W., and Robin Celikates. 2015. Towards a Critical Theory of Social Practices: Recognition, Power and Normativity. *European Journal of Philosophy* 23 (4): 839–861.

Bhandari, Anushka, and Sara Bimo. 2022. Why's Everyone on TikTok Now? The Algorithmized Self and the Future of Self-Making on Social Media. Social media + Society 8 (1).

Binder, Jens F., Andrew Howes, and Alistair Sutcliffe. 2024. The Problem of Habit in Digital Wellness and Dating Apps. *Computers in Human Behavior* 151:108025.

Brennan, Geoffrey, and Philip Pettit. 2004. *The Economy of Esteem: An Essay on Civil and Political Society*. Oxford: Oxford University Press.

Campanella, Bruno. 2022. From Mediated to Datafied Recognition: The Role of Social Media News Feeds. *Communications* 47 (4): 516–531.

Campanella, Bruno. 2024. *Recognition in the Age of Social Media*. Hoboken, NJ: John Wiley & Sons.

Cant, Callum. 2019. *Riding for Deliveroo: Resistance in the New Economy*. Cambridge: Polity Press.

Celentano, Denise, and Luigi Caranti. 2020. *Paradigms of Justice: Redistribution, Recognition, and Beyond*. Taylor & Francis.

Celikates, Robin. 2015. Digital Publics, Digital Contestation. A New Structural Transformation of the Public Sphere? In *Transformations of Democracy: Crisis, Protest, and Legitimation*, edited by Robin Celikates, Regina Kreide, and Tilo Wesche, 159–176. London: Rowman & Littlefield.

Center for Humane Technology. n.d. Time Well Spent. Accessed 21 June 2025. https://www.humanetech.com/

Coeckelbergh, Mark. 2022. *The Political Philosophy of AI: An Introduction*. Cambridge: Polity Press.

Couldry, Nick, and Andreas Hepp. 2017. *The Mediated Construction of Reality: Society, Culture, Mediatization*. Cambridge: Polity Press.

Couldry, Nick, and Ulises A. Mejias. 2019. *The Costs of Connection: How Data Is Colonizing Human Life and Appropriating It for Capitalism*. Stanford: Stanford University Press. https://doi.org/10.1515/9781503609754.

Craig, Shelley L., Kristopher M. R. McInroy, Brian Leung, and G. Nicole Krishnan. 2020. Media: A Catalyst for Resilience Among LGBTQ Youth. *Journal of LGBT Youth* 17 (3): 236–256. https://doi.org/10.1080/19361653.2019.1627709.

Davies, Will. 2021. The Politics of Recognition in the Age of Social Media. *New Left Review* 128:83–99.

Delmas, Candice. 2018. Is hacktivism the New Civil Disobedience? *Raisons Politiques* 1:63–81.

Delmas, Candice, and Kimberley Brownlee. 2021. The Philosophy of Civil Disobedience. In *Stanford Encyclopedia of Philosophy*, edited by Edward N. Zalta. Stanford: Stanford University. https://plato.stanford.edu/entries/civil-disobedience/.

Deranty, Jean-Philippe. 2009. *Beyond Communication: A Critical Study of Axel Honneth's Social Philosophy*. Leiden: Brill.

Driessens, Olivier, and Torgeir Uberg Nærland. 2022. Mediated Recognition: Identity, Respect, and Social Justice in a Changing Media Environment. *Communications* 47 (4): 505–515.

Dubber, Markus Dirk, Frank Pasquale, and Sunit Das, eds. 2020. *The Oxford Handbook of Ethics of AI*. Oxford: Oxford University Press.

Eyal, Nir. 2014. *Hooked: How to Build Habit-Forming Products*. New York: Penguin.

Eyal, Nir. 2019. *Indistractable: How to Control Your Attention and Choose Your Life*. Dallas: BenBella Books.

Fraser, Nancy. 2017. From Progressive Neoliberalism to Trump—and Beyond. *American Affairs* 1 (4): 46–64.

Fraser, Nancy, and Axel Honneth. 2003. *Umverteilung oder Anerkennung, Eine politisch-philosophische Kontroverse*. Frankfurt am Main: Suhrkamp. [2004. *Redistribution or Recognition, A Political-Philosophical Exchange*. Translated by Joel Golb, James Ingram and Christina Wilke. Verso Books].

Fricker, Miranda. 2007. *Epistemic Injustice: Power and the Ethics of Knowing*. Oxford: Oxford University Press.

Fried, Ina. 2017. *Sean Parker Unloads on Facebook: 'God Only Knows What It's Doing to Our Children's Brains'*. Axios, November 9, 2017. https://www.axios.com/2017/12/15/sean-parker-facebook-was-designed-to-exploit-human-vulnerability-1513306782.

Friedman, Milton. 1970. *The Social Responsibility of Business Is to Increase Its Profits*. New York Times Magazine, September 13, 1970.

Gerbaudo, Paolo. 2022. *The Digital Party: Political Organisation and Online Democracy*. London: Pluto Press.

Giladi, Paul. 2023. Recognition and Critical Social Epistemology. *European Journal of Philosophy* 31 (2): 412–428.

Haidt, Jonathan. 2024. *The Anxious Generation: How the Great Rewiring of Childhood Is Causing an Epidemic of Mental Illness*. New York: Penguin Press.

Haslanger, Sally. 2012. *Resisting Reality: Social Construction and Social Critique*. New York: Oxford University Press.

Hawkins, Andrew J. 2025. *How Burning Waymos Became the Image of the LA Protests*. The Verge, January 15, 2025. https://www.theverge.com/google-waymo/682932/la-protest-waymo-fire-destroy-ice-police.

Hepp, Andreas. 2020. *Deep Mediatization*. 1st ed. London: Routledge.

Hinds, Joanne, and Adam Joinson. 2018. What Demographic Attributes Do Our Digital Footprints Reveal? A Systematic Review. *PLoS One* 13 (11): e0207112. https://doi.org/10.1371/journal.pone.0207112.

Hochschild, Arlie Russell. 2016. *Strangers in Their Own Land: Anger and Mourning on the American Right*. New York: The New Press.

Honneth, Axel. 2001. Invisibility: On the Epistemology of 'Recognition'. *Proceedings of the Aristotelian Society* 75 (1): 111–126. https://doi.org/10.1111/1467-8349.00081.

Honneth, Axel. 2008. *Reification: A New Look at an Old Idea*. Oxford: Oxford University Press.

Honneth, Axel. 2010. *Das Ich im Wir: Studien zur Anerkennungstheorie*. Berlin: Suhrkamp. English translation: The I in We: Studies in the Theory of Recognition. Cambridge: Polity Press, 2012.

Honneth, Axel. [1992] 2012. *The Struggle for Recognition: The Moral Grammar of Social Conflicts*. Cambridge: Polity Press.

Honneth, Axel. 2014. *Freedom's Right: The Social Foundations of Democratic Life*. New York: Columbia University Press.

Honneth, Axel. 2018. *Anerkennung: eine europäische Ideengeschichte*. Berlin: Suhrkamp. English translation: Recognition: A European Intellectual History. Cambridge: Cambridge University Press, 2021.

Honneth, Axel. 2021. Intelligibility and Authority in Recognition: A Reply. In *Recognition and Ambivalence*, edited by Heikki Ikäheimo, Kristina Lepold, and Titus Stahl, 55–60. New York: Columbia University Press.

Hunt, Melissa G., Rachel Marx, Courtney Lipson, and Jordyn Young. 2018. No More FOMO: Limiting Social Media Decreases Loneliness and Depression. *Journal of Social and Clinical Psychology* 37 (10): 751–768.

Huq, Aziz. 2021. *The Perilous Role of Artificial Intelligence and Social Media in Mass Protests*. Modern Diplomacy, June 15, 2021. https://moderndiplomacy.eu/2021/06/15/the-perilous-role-of-artificial-intelligence-and-social-media-in-mass-protests/.

Ikäheimo, Heikki. 2017. Recognition, Identity and Subjectivity. In *The Palgrave Handbook of Critical Theory*, edited by Michael J. Thompson, 567–585. London: Palgrave Macmillan.

Ikäheimo, Heikki, and Arto Laitinen. 2007. Analyzing Recognition: Identification, Acknowledgement, and Recognitive Attitudes towards Persons. In *Recognition and Power*, edited by Bert van den Brink and David Owen, 33–56. Cambridge: Cambridge University Press.

Ikäheimo, Heikki, Kristina Lepold, and Titus Stahl, eds. 2021. *Recognition and Ambivalence*. New York: Columbia University Press.

Iser, Mattias. 2008. *Empörung und Fortschritt: Grundlagen einer kritischen Theorie der Gesellschaft*. Frankfurt am Main: Campus Verlag.

Iser, Mattias. 2019. Recognition. In *Stanford Encyclopedia of Philosophy*, edited by Edward N. Zalta. Stanford: Stanford University. https://plato.stanford.edu/entries/recognition/.

Jackson, Sarah J., Moya Bailey, and Brooke Foucault Welles. 2020. *#HashtagActivism: Networks of Race and Gender Justice*. Cambridge, MA: MIT Press.

Jacobsen, B. N. 2021. Regimes of Recognition on Algorithmic Media. *New Media and Society* 25 (12): 3641–3656. https://doi.org/10.1177/14614448211053555. (Original work published 2023).

Jansson, André. 2015. Interveillance: A New Culture of Recognition and Mediatization. *Media and Communication (Lisboa)* 3 (3): 81–90. https://doi.org/10.17645/mac.v3i3.305.

Khalil, Joe. 2020. *Social Media Algorithms Boost L.A. Protest Misinformation in 'Combustible' Environment*. CNN Business, June 1, 2020. https://www.cnn.com/2020/06/01/tech/social-media-misinformation-george-floyd-protests/index.html.

Koenig, Biko, and Tali Mendelberg. 2025. The Symbolic Politics of Status in the MAGA Movement. *Perspectives on Politics*1–18.

Kormelink, T. G., F. Houwing, B. Struminskaya, L. Boeschoten, N. de Schipper, and K. Welbers. 2025. Meaningful Informed Consent? How Participants Experience and Understand Data Donation. *Information, Communication and Society* 28 (1): 1–18. https://doi.org/10.1080/1369118X.2025.2540915.

Laitinen, Arto, and Otto Sahlgren. 2021. AI Systems and Respect for Human Autonomy. *Frontiers in Artificial Intelligence* 4:151.

Lepold, Kristina. 2019. Examining Honneth's Positive Theory of Recognition. *Critical Horizons* 20 (3): 246–261.

Lewis, Paul. 2017. *'Our Minds Can Be Hijacked': The Tech Insiders Who Fear a Smartphone Dystopia*. The Guardian, October 6, 2017. https://www.theguardian.com/technology/2017/oct/05/smartphone-addiction-silicon-valley-dystopia.

Magalhães, João C., and Jun Yu. 2022. Social Media, Social Unfreedom. *Communications* 47 (4): 553–571.

Markell, Patchen. 2003. *Bound by Recognition*. Princeton: Princeton University Press.

McInroy, Kristopher M. R., and Shelley L. Craig. 2017. You Can Form a Part of Yourself Online: The Influence of New Media on Identity Development and Coming Out for LGBTQ Youth. *Journal of Gay and Lesbian Mental Health* 21 (2): 95–109. https://doi.org/10.1080/19359705.2016.1267352.

Medina, José. 2018. Epistemic Injustice and Epistemologies of Ignorance. In *Routledge Handbook of Epistemic Injustice*, edited by Ian James Kidd, José Medina, and Gaile Pohlhaus Jr., 247–258. London: Routledge.

Medina, José. 2023. *The Epistemology of Protest: Silencing, Epistemic Activism, and the Communicative Life of Resistance*. Oxford: Oxford University Press.

Metzler, Hannah, and David Garcia. 2024. Social Drivers and Algorithmic Mechanisms on Digital Media. *Perspectives on Psychological Science* 19 (4): 735–748. https://doi.org/10.1177/17456916231185057.

Milan, Stefania. 2013. *Social Movements and Their Technologies: Wiring Social Change*. Basingstoke: Palgrave Macmillan.
Mills, Charles W. 2018. Black Radical Kantianism. *Res Philosophica* 95 (1): 1–33.
Moody-Adams, Michele M. 2022. *Making Space for Justice: Social Movements, Collective Imagination, and Political Hope*. New York: Columbia University Press.
Naaman, Mor, Jeffrey Boase, and Chih-Hui Lai. 2010. Is it Really About Me? Message Content in Social Awareness Streams. In *Proceedings of the 2010 ACM Conference on Computer Supported Cooperative Work*, 189–192.
Noble, Safiya Umoja. 2018. *Algorithms of Oppression: How Search Engines Reinforce Racism*. New York: NYU Press.
O'Neil, Cathy. 2022. *Weapons of Math Destruction: How Big Data Increases Inequality and Threatens Democracy. Updated Edition*. New York: Crown.
Origgi, Gloria, and Serena Ciranna. 2017. Epistemic Injustice: The Case of Digital Environments. In *The Routledge Handbook of Epistemic Injustice*, 303–312. Routledge.
Peters, John Durham. 1999. *Speaking into the Air: A History of the Idea of Communication*. Chicago: University of Chicago Press.
Pollmann, Arnd. 2008. Recognition as the Foundation of Morality and Law. In *Philosophical Foundations of Human Rights*, edited by Rowan Cruft, S. Matthew Liao, and Massimo Renzo, 317–330. Oxford: Oxford University Press.
Polyakova, Alina, and Chris Meserole. 2019. How Autocrats Weaponize AI—And How to Fight Back. *Journal of Democracy* 30 (1): 48–55.
Renault, Emmanuel. 2019. *The Experience of Injustice: A Theory of Recognition*. Columbia University Press.
Reuters. 2021. *Signal Downloads Surge After WhatsApp Privacy Update Backlash*. January 12, 2021. https://www.reuters.com/article/us-facebook-whatsapp-signal-idUSKBN29H1DB.
Rideout, Victoria, Alanna Peebles, Supreet Mann, and Michael B. Robb. 2022. *The Common Sense Census: Media Use by Tweens and Teens, 2021*. San Francisco: Common Sense Media.
Roessler, Beate. 2020. Privacy and Autonomy: On Some Misconceptions Concerning the Political Dimensions of Privacy. *European Journal of Political Theory* 19 (2): 193–213.
Rosenberg, Matthew, Nicholas Confessore, and Carole Cadwalladr. 2018. *How Trump Consultants Exploited the Facebook Data of Millions*. The New York Times, March 17, 2018.
Rosenblat, Alex. 2018. *Uberland: How Algorithms Are Rewriting the Rules of Work*. Berkeley: University of California Press.
Shahbaz, Adrian. 2021. How Autocrats Weaponize AI—And How to Fight Back. *Journal of Democracy*, online exclusive. https://www.journalofdemocracy.org/online-exclusive/how-autocrats-weaponize-ai-and-how-to-fight-back/.

Snow, Jacob. 2018. *Amazon's Face Recognition Falsely Matched 28 Members of Congress With Mugshots*. ACLU News & Commentary, July 26, 2018. https://www.aclu.org/news/privacy-technology/amazons-face-recognition-falsely-matched-28.

Susskind, Jamie. 2018. *Future Politics: Living Together in a World Transformed by Tech*. Oxford: Oxford University Press.

Syvertsen, Trine. 2020. *Digital Detox: The Politics of Disconnecting*. Bingley: Emerald Publishing.

Taylor, Charles. 1994. The Politics of Recognition. In *Multiculturalism: Examining the Politics of Recognition*, edited by Amy Gutmann, 25–73. Princeton: Princeton University Press.

Thompson, Simon. 2006. *The Political Theory of Recognition: A Critical Introduction*. Cambridge: Polity Press.

Tiffany, Kaitlyn. 2019. *Online Ads Can Be Targeted Based on Your Emotions*. Vox, May 21, 2019. https://www.vox.com/the-goods/2019/5/21/18634323/new-york-times-emotion-based-ad-targeting-sadness.

Tilly, Charles, and Sidney Tarrow. 2015. *Contentious Politics*. 2nd ed. Oxford: Oxford University Press.

Tucker, Joshua A., Andrew Guess, Pablo Barberá, Cristian Vaccari, Alexandra Siegel, Sergey Sanovich, Denis Stukal, and Brendan Nyhan. 2018. Social Media, Political Polarization, and Political Disinformation: A Review of the Scientific Literature. *Political Communication* 35 (4): 503–533.

Véliz, Carissa. 2021. *Privacy Is Power: Why and How You Should Take Back Control of Your Data*. London: Bantam.

Venkatesh, Nikhil. 2021. Surveillance Capitalism: A Marx-Inspired Account. *Philosophy* 96 (3): 359–385.

Waelen, Rosalie. 2022. The Struggle for Recognition in the Age of Facial Recognition Technology. *AI and Ethics* 2 (3): 1–8.

Zuboff, Shoshana. 2019. *The Age of Surveillance Capitalism: The Fight for a Human Future at the New Edge of Power*. New York: Public Affairs.

Zurn, Christopher F. 2015. *Axel Honneth: A Critical Theory of the Social*. Cambridge: Polity Press.

Zwolinski, Matt, Alan Wertheimer, and Ruth Sample. 2022. Exploitation. In *Stanford Encyclopedia of Philosophy*, edited by Edward N. Zalta. Stanford: Stanford University. https://plato.stanford.edu/entries/exploitation/.

Struggles for Recognition in the Digitalizing Workplace

Joel Anderson

Abstract This chapter argues that Axel Honneth's recognition theory provides key resources for diagnosing what makes AI-mediated labor normatively troubling. Efficiency-focused frameworks miss the unease that is increasiingly shaping the political landscape of these rapidly adopted technologies: eroded collegial bonds, blurred sense of achievement, and the strangeness of "recognition" from algorithms. I examine three transformed aspects of professional work (coordination, achievement, and performance assessment) across their constitutive, the causal-psychological, and the normative-institutional dimensions. I introduce the term **synthetic recognition** in discussing a distinctive pathological characteristic of algorithmic evaluations: that their simulated acknowledgment (and possible psychological benefits) emerge in unidirectional feedback from an invulnerable system that violates the defining qualities of genuine recognition's reciprocity and mutual vulnerability. By drawing on Honneth's concept of *social freedom* and the link he makes (in *The

Work on this chapter was partly funded by the research programme Ethics of Socially Disruptive Technologies, which is funded by the Gravitation programme of the Dutch Ministry of Education, Culture, and Science and the Netherlands Organization for Scientific Research (NWO grant number 024.004.031).

J. Anderson (✉)
Department of Philosophy and Religious Studies, Utrecht University, Utrecht, The Netherlands
e-mail: j.h.anderson@uu.nl

R. Waelen (ed.), *Normative Relations of Recognition in the Age of AI*, Philosophy and Ethics of AI 2, https://doi.org/10.1007/978-3-032-34028-3_3

Working Sovereign*) between workplace democracy and democratic citizenship – and by supplementing this with insights from relational autonomy and democratic-equality theory, I argue that recognition-sensitive AI requires democratic authorization, not merely better design.

Keywords Recognition • Digitalization • Future of work • Automation • Identity • Social reproduction

1 Introduction

The integration of artificial intelligence technologies into contemporary workplaces represents far more than a technical evolution in productive capacity. Unlike previous waves of automation that primarily displaced manual labor through mechanical substitution, AI's penetration into cognitive and communicative tasks fundamentally challenges the normative foundations upon which professional identity and workplace recognition have traditionally been grounded. When AI systems diagnose medical conditions, generate legal briefs, coordinate complex projects, or evaluate worker performance, they do not merely augment human capabilities; they reconfigure the very structures through which workers understand themselves as making meaningful contributions to social reproduction and thereby deserving of esteem and recognition.

Consider the cases that are becoming increasingly familiar: the professional worker whose daily tasks are increasingly mediated by artificial intelligence systems; a financial analyst using machine learning algorithms to identify investment patterns that would take weeks to discover manually; a radiologist who relies on AI-assisted diagnostic tools that flag potential abnormalities with unprecedented accuracy; or a software developer who employs code generation assistants that produce complex functions from simple natural language prompts. In each case, productivity soars, quality improves, and efficiency metrics climb. Yet these same professionals report a peculiar unease, a sense that something normatively significant about their work has shifted, even as objective performance indicators suggest nothing but gains.

This paradox (of technological enhancement accompanied by experiential impoverishment and deep normative disease) resists explanation through conventional analytical frameworks. Standard accounts of workplace automation focus on quantitative concerns: job displacement, wage effects, skill requirements, economic distribution. (On socially and conceptually disruptive technologies more generally, see Van de Poel et al. 2023.) While these material considerations matter profoundly, they miss what workers themselves articulate when reflecting on AI-mediated labor: concerns about authentic professional competence, erosion of collegial bonds, the strange unreality of receiving performance feedback from algorithmic systems, the difficulty of claiming ownership over achievements to which AI contributed substantially. These experiential dimensions suggest that something more than economic welfare is at stake in AI's workplace integration.

This chapter draws on earlier work (Anderson 2025b) and especially resources from critical social theory, particularly Axel Honneth's recognition theory and its roots in the Frankfurt School tradition, to illuminate these internal, "subjective" concerns about AI-mediated labor. Recognition theory identifies work as a crucial sphere for moral experience and social esteem in modern societies (Honneth 2008a, 2008b, 2023, 2024a, 2024b). Through work, individuals gain not merely livelihood for themselves and their dependents but also self-respect, identity, and integration into social communities. When AI systems transform work's organization and experience, they potentially disrupt these recognition relations in ways that threaten both individual flourishing and democratic capacity. Understanding these disruptions requires theoretical resources capable of analyzing work not merely as economic activity but as a sphere of mutual recognition essential to human agency and social freedom.

This chapter contributes to the emerging discussion of critical social theory and AI technologies by examining the sphere of labor, a sphere Honneth identifies as particularly crucial for modern identity formation, yet one that has received limited attention in philosophical discussions of AI. While AI affects many aspects of our lives, and of our recognition relations, work occupies a distinctive mediating position between personal development and social contribution. For most adults, work consumes the majority of waking hours and provides the primary context for exercising capabilities, contributing to collective purposes, and experiencing recognition from others. AI's transformation of this sphere therefore carries implications beyond individual welfare, extending to the social conditions that support both personal autonomy and democratic citizenship.

A clarification is necessary here regarding the baseline against which AI's transformative impact is assessed. I am not claiming that pre-AI workplaces were recognition-rich environments subsequently degraded by technology. The history of capitalist labor relations is, as Marx, Durkheim, and Honneth all acknowledge, a history of struggle against alienation, exploitation, and misrecognition. Many workers have never experienced workplaces as sites of genuine mutual acknowledgment.

The concern, rather, is twofold. First, AI threatens to foreclose possibilities for recognition that remained available even in imperfect workplaces—the potential for collegial bonds, the space for contesting unfair evaluations, the developmental pathways through which workers might cultivate autonomous judgment. Second, and more subtly, AI systems often eliminate the visibility of domination. When a human supervisor exercises arbitrary power, workers can at least identify the source of domination and potentially organize against it. When algorithmic systems coordinate, evaluate, and discipline, the locus of control becomes diffuse, opaque, and difficult to contest. The comparison, then, is not between an idealized past and a degraded present, but between forms of workplace organization that differ in their susceptibility to democratic transformation.

The argument proceeds in four stages. First (Sect. 2), I reconstruct recognition theory as an analytical framework, tracing its development from Marx's critique of alienated labor and Durkheim's analysis of organic solidarity through Honneth's identification of work as a sphere of mutual recognition supporting practical relations-to-self constitutive of autonomous agency. This establishes work's normative significance beyond instrumental productivity.

The chapter then examines three aspects of professional work being transformed by AI integration (Sects. 3, 4, and 5): coordination and collaboration, work quality and efficiency, and performance assessment. For each aspect, I analyze AI's "transformative impact on recognition relations" across three dimensions drawn from critical theory: the constitutive (how AI changes what work *is*), the causal-psychological (how AI affects workers' capacities for autonomous engagement), and the normative-institutional (how AI transforms legitimacy and governance of workplace relations). In addition, I introduce the concept of synthetic recognition to capture the distinctive pathology of algorithmic performance assessment, where evaluative feedback simulates genuine recognition while lacking its essential intersubjective structure.

Finally (Sect. 6), I offer concluding remarks that synthesize key findings while distinguishing areas of greater concern from those offering more opportunities for mitigation or even positive transformation. Throughout, I emphasize recognition theory's distinctive contribution to understanding what makes AI-mediated labor troubling even when it increases productivity, a contribution that efficiency-focused frameworks systematically miss.

2 Recognition Theory as a Normative Framework: Work, Alienation, and Social Freedom

2.1 Marx's Critique of Alienated Labor and Its Legacy

The philosophical foundations for analyzing AI-mediated work begin with Karl Marx's early critique of "estranged" or "alienated" labor in his 1844 Economic and Philosophical Manuscripts (Marx 1988). Marx identified four dimensions through which workers become alienated under capitalist production: from the products of their labor (unable to recognize themselves in what they produce), from the production process itself (activity becomes mere means to survival rather than creative expression), from their "species-being" (distinctively human creative potential), and crucially, from fellow workers (Marx 2007).

This final dimension, social alienation, proves particularly significant for understanding AI's workplace impact. As Marx observed, "A direct consequence of man's alienation from the product of his work, from his life activity, and from his species-existence, is the alienation of man from man." (Marx 1988, Section XIV). When workers cannot recognize themselves in products or processes, they simultaneously lose the intersubjective context in which mutual recognition becomes possible. Work ceases to be a sphere of cooperative self-realization and becomes instead isolated instrumental activity.

Marx's insight that alienation is fundamentally *social* rather than merely individual or economic anticipates recognition theory's central concern. The problem isn't simply that workers feel dissatisfied or earn insufficient wages (though these matter) but that the structure of alienated labor prevents the mutual acknowledgment through which subjects develop and sustain autonomous agency. When AI systems enable increasingly isolated task completion or obscure individual contributions through opaque collaboration, they risk recreating this Marxian alienation in novel technological forms.

Contemporary critical theorists have extended and refined Marx's analysis. Rahel Jaeggi reconstructs alienation as "a relation of relationlessness"—not the absence of relation to world and self but a deficient one preventing appropriation (Jaeggi 2014). Her four dimensions map precisely onto AI-mediated work's pathologies: algorithmic management creates powerlessness before opaque systems, platform design atomizes workers producing indifference, micro-tasking fragments meaningful activity, and surveillance capitalism demands self-instrumentalization. Crucially, Jaeggi distinguishes alienation from objectification; not all externalization of human capacities in technical systems constitutes alienation, only those forms where objectified structures dominate subjects (Jaeggi 2014). This allows a more nuanced and specific analysis of when AI enables versus obstructs human appropriation of work processes.

2.2 Durkheim on Division of Labor and Organic Solidarity

In a way particularly relevant to the recognition-theoretic approach to labor, Émile Durkheim's *The Division of Labor in Society* (1984) complements Marx's analysis by emphasizing work's role in generating social cohesion. Durkheim argued that specialized roles create mutual interdependence and hence "organic solidarity," social bonds emerging not despite but *because of* differentiation. Unlike the "mechanical solidarity" of traditional societies based on sameness, modern social cohesion arises through recognizing how diverse, specialized contributions form an interdependent whole.

This organic solidarity carries normative significance beyond functional efficiency. Workers don't merely coordinate instrumentally; they come to see themselves as participants in a collective enterprise whose success depends on each person's distinctive contribution. Through this interdependence, individuals achieve recognition, that is, acknowledgment of their particular capabilities and their meaningful role in social reproduction. The division of labor thus functions not only economically but also morally, generating bonds of reciprocity and shared purpose.

Durkheim's framework provides important conceptual tools for illuminating specific pathologies of AI-mediated work. When algorithmic systems coordinate tasks without workers understanding their place in larger processes, or when platform architecture fragments contributions into microtasks invisible to others, the conditions for organic solidarity erode. Workers may complete assigned tasks efficiently yet never experience the mutual recognition that comes from understanding how their distinctive contribution meshes with others' in accomplishing shared objectives. The division of labor persists but loses its solidarity-generating function, leaving isolated individuals coordinating instrumentally rather than cooperating meaningfully.

This concern extends beyond subjective dissatisfaction to threaten social integration itself. If work no longer generates organic solidarity, becoming mere instrumental coordination among atomized individuals, society loses a crucial mechanism for cultivating reciprocity and shared purpose. AI systems that optimize task distribution for efficiency while obscuring interdependencies risk undermining the moral foundations Durkheim identified as essential for modern social cohesion.

2.3 Honneth's Recognition Theory: Three Spheres and Work's Role

More than perhaps any other critical approach within the Frankfurt School tradition, Axel Honneth's recognition theory provides a comprehensive framework for analyzing work's normative significance in contemporary societies. In *The Struggle for Recognition* (1995), Honneth identified three spheres where subjects gain distinct forms of recognition necessary for autonomous agency: love (emotional support grounding basic self-confidence or self-trust), rights (legal respect grounding self-respect), and solidarity (social esteem grounding self-esteem). Each sphere contributes to subjects' "positive relation-to-self," that is, the practical self-understanding enabling them to act as autonomous agents.

Initially, Honneth analyzed work primarily through the solidarity sphere, where individuals achieve recognition for contributions society values (Honneth 1995). Responding to challenges of the adequacy of his concept of work and recommendations to rehabilitate his early, more Marxian account (for an overview, see Boston 2018), he subsequently developed more sophisticated accounts of work's multiple recognition functions. In "Arbeit und Anerkennung [Labor and Recognition]" (2008), Honneth argues work enables recognition not through intrinsic qualities but instrumentally—by providing economic security for intimate relationships (love sphere), workplace rights and dignity (legal sphere), and social esteem for contributions (solidarity sphere). This "redetermination" avoids both utopian perfectionism (work as self-realization) and reductive economism (work as mere survival), instead evaluating work through its role in enabling recognition across all three spheres.

In Honneth's more recent work, particularly *The Working Sovereign* (2024a), transforms this analysis by grounding it in democratic theory rather than perfectionist ideals about labor's intrinsic value. His central thesis is that democratic participation requires a fair and transparent division of labor, with workplaces functioning as "rehearsal spaces" for democratic interaction (Honneth 2024a, 2024b). He identifies five dimensions through which work either enables or undermines democratic citizenship: economic independence, temporal autonomy for civic engagement, psychological conditions of self-respect, social recognition preventing subordination, and material conditions preventing exhaustion (Honneth 2023; 2024a, 2024b).

This five-dimensional framework proves crucial for analyzing AI-mediated labor. The economic dimension concerns whether work provides sufficient income for political engagement without precarity anxiety. The temporal dimension addresses whether work schedules allow civic participation. The psychological dimension examines whether workplace relations support or undermine self-respect. The social dimension analyzes whether workers experience recognition preventing subordination. The material dimension considers whether work exhausts workers physically and mentally, foreclosing political thought (Honneth 2023).

For Honneth, the relationship between work and democracy is constitutive, not merely instrumental. Workers who experience authoritarianism, surveillance, and disrespect at work cannot fully develop capacities for democratic citizenship (Honneth 2024a, 2024b; see also Celikates, Honneth, and Jaeggi 2023). This "spill-over thesis" means workplace conditions directly shape citizens' capacity for political participation. If AI systems transform work in ways that undermine recognition across these five dimensions, they threaten not merely individual welfare but the social conditions necessary for democratic life itself.

2.4 Social Freedom and Mutual Recognition

Central to Honneth's analysis is the concept of "social freedom," freedom actualized through participation in social relationships and institutions (Honneth 2014). Unlike negative freedom (absence of interference) or reflective freedom (rational self-determination), social freedom requires both subjective self-determination and objective social conditions enabling its realization. Work represents a paradigmatic sphere where, when properly structured, individuals exercise capabilities, contribute to collective purposes, and experience recognition—actualizing freedom through social cooperation (Honneth 2014).

In an incisive parallel discussion of the university, O'Neill and Smith argue social freedom requires recognized participation in three spheres: personal relationships, productive work, and democratic citizenship (O'Neill and Smith 2022). Workplace institutions should promote social freedom by enabling workers to realize individual autonomy within social roles and relationships, strengthening democratic capacities.

AI's distinctive challenge is that it threatens social freedom by transforming the cooperative structures through which individuals actualize freedom together. When algorithmic management replaces collaborative coordination, when synthetic recognition substitutes for genuine interpersonal acknowledgment, when opaque systems mediate all workplace relations, the institutional conditions for social freedom erode.

2.5 Complementary Frameworks: Relational Autonomy and Democratic Equality

Two additional frameworks complement recognition theory's resources for critical diagnosis. First, relational autonomy theory challenges traditional individualistic autonomy conceptions by showing how personal autonomy is fundamentally social and relational (Mackenzie and Stoljar 2000). Catriona Mackenzie develops a three-dimensional analysis distinguishing self-determination, self-governance, and self-authorization as distinct but interacting dimensions. Self-authorization (believing one has normative authority to be self-determining and self-governing) is constitutively social, formed within normative practices of mutual recognition (Mackenzie 2014).

This framework proves crucial for understanding how AI-mediated work environments affect workers' relational autonomy. When AI systems mediate workplace relationships and structure opportunities for autonomous action, they function as "social scaffolding" either enabling or constraining workers' self-determination and self-authorization. Algorithmic management that denies workers' capacity for discretionary judgment undermines the social conditions necessary for autonomous agency itself (Anderson 2025a).

Second, Elizabeth Anderson's democratic equality framework provides comprehensive criteria for evaluating AI work: do systems enable workers to function as equals in three spheres: as human beings, participants in cooperative production, and democratic citizens? (Anderson 1999, 2017). Her critique of workplace authoritarianism ("private government?") illuminates how algorithmic control constitutes domination even when not actively oppressive, since arbitrary power over workers' lives violates republican freedom (Anderson 2017). Workplaces are "private governments" with sweeping arbitrary power that would be unconstitutional if wielded by the state. This framework shows why lack of democratic control over AI governance represents domination requiring democratic accountability.

Together, these frameworks establish work's normative significance across multiple dimensions: as a sphere of mutual recognition essential for autonomous agency (Honneth), as social scaffolding for relational autonomy (Mackenzie), and as a site where democratic equality is either realized or violated (Anderson). AI's transformation of work therefore carries stakes beyond efficiency or economic welfare—it affects the social conditions supporting both personal flourishing and democratic capacity.

Before turning to the three domains of AI-transformed work, a methodological note is in order regarding the analytical framework that is at work, albeit largely in the background. In analysing each domain, I distinguish three dimensions of AI's impact on recognition relations: the constitutive, the causal-psychological, and the normative-institutional. These correspond to three distinct questions about the relationship between work and autonomous agency, questions that are frequently conflated in discussions of technology and labour, generating avoidable confusion.

The constitutive dimension asks what a given aspect of work *is*, that is, what structures of mutual recognition it presupposes. This is a claim about the nature of workplace relations: that certain forms of authority, competence, and collegial standing exist only within spaces of mutual recognition and cannot be understood as properties of isolated individuals. When AI transforms these structures, it alters not merely the conditions under which workers exercise autonomy but also the very architecture within which autonomous agency is possible.

The causal-psychological dimension asks how AI affects workers' (onboard) capacities for autonomous engagement. This is an empirical question about what happens when the social practices through which workers develop judgment, self-trust, and collaborative skill are disrupted or foreclosed. These are testable claims, best adjudicated by behavioral and social scientists who examine what occurs when dialogical competence develops without actual dialogue, or when resilience is never built through recognised successes.

The normative-institutional dimension asks whether AI-mediated workplace arrangements are legitimate, that is, whether they are authorised by those whose autonomy they affect. The inclusion of this dimension is what prevents the analysis from collapsing into either metaphysical speculation or empirical psychology: it generates concrete institutional demands.

These three dimensions are heuristic rather than exhaustive, and they are empirically intertwined. But distinguishing them prevents a characteristic error in debates about AI and work: criticising the constitutive claim with arguments that address only causal effects, or defending workplace AI against paternalism objections while ignoring the normative requirements for legitimate scaffolding. I now examine three aspects of professional work being transformed by AI—coordination (Sect. 3), work quality and achievement

(Sect. 4), and performance assessment (Sect. 5)—through each of these three lenses.

3 Coordination and Collaboration: From Organic Solidarity to Algorithmic Fragmentation

3.1 The Constitutive Dimension: What Work Becomes

The first of three aspects of work being transformed by AI technologies I will consider is the the impact on coordination. Pre-AI professional work typically involved significant collaborative coordination—team members planning projects together, negotiating role divisions, communicating progress, adjusting contributions in response to others' work. This coordination wasn't merely instrumental but generated the organic solidarity Durkheim identified: through coordinating, workers came to understand their interdependence and developed bonds of reciprocity. Collaborative work functioned as a site of mutual recognition where colleagues acknowledged each other's distinctive contributions to shared purposes.

AI coordination tools fundamentally transform this structure. Project management systems automatically assign tasks based on algorithmic assessment of skills, availability, and priority. Workflow automation routes documents and approvals without human deliberation. Communication platforms structure interactions through designed affordances that shape what coordination looks like. Recommendation systems suggest collaborators based on pattern matching rather than workers' judgment about whose expertise would contribute meaningfully.

This transformation has both efficiency benefits and recognition costs. Algorithmically-mediated coordination often proceeds more smoothly than traditional negotiation; tasks get assigned, handoffs occur seamlessly, bottlenecks are identified automatically. Yet something normatively significant changes: workers coordinate instrumentally without the intersubjective encounter through which colleagues recognize each other as cooperating agents. The division of labor persists, but the solidarity-generating process Durkheim described is dramatically diminished.

In a recent article, Juri Viehoff (2025) has sharpened this point by distinguishing *productive solidarity* (solidarity in joint creation) from *workplace-solidarity* (solidarity in shared structural position). AI-mediated, "colleague-less" settings threaten both: they thin out the communicative practices that let contributors recognize one another's roles, and they isolate workers in ways that make collective contestation harder to organize. On his view, the imperative is not AI-abstinence but solidarity-preserving AI, tools that scaffold cooperation rather than substitute for it.

Consider a software development team using AI-enhanced project management. Previously, even "sprint" or "agile" planning involved collective deliberation: developers discussed feature complexity, negotiated task assignments, identified interdependencies, agreed on priorities. Without romanticizing it, the problem-solving task constitutive of this deliberative process is at least oriented towards finding mutual understanding of how each person's work fitted into the larger project and why their distinctive contribution mattered. With AI task assignment optimizing for efficiency metrics, this deliberative space contracts. Developers receive assignments, complete them, and move to the next task, coordinating effectively but losing the collaborative encounter that generated organic solidarity.

The constitutive transformation is subtle but significant: work remains cooperative in structure yet loses the intersubjective dimension that made cooperation recognizable as such. Workers are coordinated but may not experience themselves as collaborating. They contribute to collective outputs yet lack clear understanding of how their distinctive work meshes with others' or why it matters to shared purposes.

3.2 The Causal-Psychological Dimension: Erosion of Relational Autonomy Skills

The psychological impact of this transformation extends beyond subjective dissatisfaction to affect workers' capacity for autonomous engagement. Collaborative coordination serves crucial developmental functions: through negotiating with colleagues, workers develop judgment about appropriate task division; through explaining their work to others, they articulate its purpose and meaning; through adjusting contributions in response to feedback, they cultivate practical wisdom about effective cooperation.

When AI systems mediate these processes, these developmental opportunities contract. Workers may become skilled at following algorithmic assignments and completing isolated tasks yet lose capacity for the autonomous judgment that collaborative coordination cultivates. The result is an erosion of the relational scaffolding supporting autonomous agency and, more broadly, the social context where workers develop self-authorization through recognized participation in cooperative projects.

This is actually a good example of something Mackenzie's relational autonomy framework illuminates: autonomy requires social scaffolding, and algorithmic coordination can undermine the relational context necessary for autonomous capacity development (Anderson 2024; Mackenzie 2014). Workers may retain formal authority to make decisions yet lack the practical judgment that emerges from collaborative experience. They can complete assigned tasks but struggle to initiate meaningful projects requiring coordination of multiple contributors.

Empirical research documents these psychological effects. Studies of algorithmic management in platform work reveal how information asymmetries—where platforms possess comprehensive data while workers confront opaque algorithmic bosses—systematically undermine workers' capacity for autonomous judgment (Rosenblat and Stark 2016; Christiaens 2022. Uber drivers report experiencing the algorithm as an "invisible boss" that directs behavior while denying them understanding of the logic governing assignments. This creates an illusion of self-determination; workers nominally control their schedule yet effectively respond to algorithmic nudges they cannot fully comprehend or resist.

The long-term developmental concern is whether workers subjected to extensive algorithmic coordination retain capacity for the autonomous collaborative judgment that historically characterized professional work. If algorithms handle coordination, do workers lose not merely current opportunities for collaboration but also the developmental pathway through which they would cultivate autonomous coordination capacity?

3.3 The Normative-Institutional Dimension: Democracy and Workplace Governance

The question of institutional legitimacy concerns who controls coordination tools and for what purposes. In traditional workplaces, coordination methods evolved through collective practice: teams developed conventions, negotiated procedures, adjusted approaches based on experience. While managers retained ultimate authority, workers possessed significant voice in shaping coordination practices—and if they didn't, they could fight for it.

Algorithmic coordination transforms this dynamic by encoding coordination logic into the systems designed by technical teams responding to management directives. Workers may have no voice in determining how tasks get assigned, what metrics drive assignments, which factors the algorithm considers relevant, or how the system evolves. This represents a clear instance of structural domination, in the sense in which workers are subject to arbitrary power over their choices.

The normative concern isn't merely that algorithms coordinate differently than humans but that their deployment occurs without democratic authorization from those most affected. Workers are subjected to algorithmic coordination systems they did not design, cannot fully understand, and have limited capacity to contest. This violates basic democratic principles that those subject to governing systems should have meaningful voice in their design and operation.

Honneth's framework makes this concern more precise. On his view, workplace democracy is not merely desirable but constitutive for political democracy (Honneth 2023, 2024a, 2024b). When workers experience unilateral, uncontestable algorithmic control at work, they learn subordination rather than democratic participation. The workplace ceases functioning as a "rehearsal space" for democratic interaction and becomes instead a site where authoritarian governance is normalized. This threatens not merely workplace dignity but democratic capacity itself—and contributes to a further dimension of deskilling, in the domain of practices of critical self-determination.

What is threatened, then, is democratic governance of coordination technologies. This doesn't require rejecting algorithmic tools entirely but rather subjecting their design and deployment to meaningful worker participation. Various models exist: co-determination systems where workers have formal voice in technology decisions, platform cooperatives where workers collectively own and govern coordination systems, participatory design processes where users shape tool development. The crucial requirement is transforming algorithmic coordination from unilateral management prerogative to collectively governed infrastructure serving shared purposes.

Recent work on workplace democracy and algorithmic governance underscores on this point. Roberto Frega, Lisa Herzog, and Christian Neuhäuser argue workplace democracy is needed for meaningful work, relational equality, and avoiding domination (Frega et al. 2019). Onni Hirvonen and Keith Breen demonstrate how authoritarian workplaces—whether managed by humans or algorithms—systematically deny all three recognition forms (Hirvonen and Breen 2020). Whatever the potential may be for new forms of critical engagement with over the work of coordination, there are profound challenges in this domains to keep the coordination aspect of work meaningful. Especially given the widespread and understandable antipathy for meetings, the idea of handing coordination tasks off to the AI has an enormous appeal. But this may well be a case in which we should be careful what we wish for.

4 Work Quality and Efficiency: The Achievement Gap and Attribution Challenges

4.1 The Constitutive Dimension: Blurred Boundaries of Achievement

I turn now to a second aspect of work affected by AI: the significant (though certainly not universally realized) potential for improvements in the quality and efficiency of the tasks completed. AI systems increasingly participate in core professional tasks previously requiring distinctive human expertise. Legal AI assists case research and document drafting. Medical AI supports diagnostic interpretation. Creative AI generates text, images, and code. Engineering AI optimizes designs and simulates performance. In each domain, AI tools enhance human capabilities, enabling faster analysis, identifying patterns humans would miss, generating options humans might not consider.

This enhancement creates what John Danaher and Sven Nyholm call the "achievement gap"—as AI systems automate skilled tasks, they eliminate opportunities for genuine human achievement requiring causal efficacy, appropriate difficulty, relevant control, and meaningful outcomes (Danaher and Nyholm 2021). The achievement gap represents the inverse of the responsibility gap: while responsibility gaps ask who's blamed when AI fails, achievement gaps ask who's credited when AI succeeds.

One might object that AI collaboration will generate new forms of recognition adapted to hybrid achievement. Just as the Romantic conception of solitary artistic genius gave way to recognition of collaborative and technically-mediated creativity, perhaps professional recognition will evolve to esteem effective AI orchestration, prompt engineering, or human-machine coordination skills.

This response is worth taking seriously. It is possible that cultural norms will shift to recognize new forms of expertise grounded in effective AI collaboration rather than AI-independent task performance. However, two concerns remain. First, such a shift requires substantial cultural and institutional transformation that cannot be assumed—current reward structures, professional identities, and recognition practices remain oriented toward individual achievement, creating a transitional period where workers experience genuine recognition deficits. Second, and more fundamentally, some dimensions of recognition may not be transferable to hybrid contexts. The self-esteem that comes from confronting a difficult challenge and succeeding through one's own developed capabilities has a phenomenological character that "I prompted the AI effectively" may not replicate. This is not technological conservatism but rather an observation that different forms of achievement ground different forms of self-relation, and not all substitutions are equivalent from the perspective of human flourishing.

The problem isn't that work becomes easier but that achievement boundaries blur. When a lawyer uses AI-assisted research, is the resulting brief their achievement, the AI's, or some irreducible combination? When a data scientist's AI model yields insights, should they receive credit proportional to their direct contribution (data preparation, parameter tuning) or for the complete outcome including the AI's pattern recognition? When a graphic designer employs generative AI, does the creative achievement belong to them, the AI developers who trained the model, or the countless artists whose work the AI learned from? This is usefully understood as a conceptual disruption: a challenge to the settled application conditions of a concept (Löhr 2023; Hopster 2024).

These aren't merely philosophical puzzles but lived concerns affecting professional identity and self-esteem. Professional identity historically centered on mastery of distinctive skills, the capacity to accomplish what others cannot. When AI enables novices to match experts, this identity foundation erodes. Research on AI augmentation tools reveals tensions between productivity gains and potential deskilling that undermines professional identity and peer recognition. The "leveling effect" (where AI enables novices to match experts) may democratize access but simultaneously devalues the expertise grounding professional recognition (Danaher and Nyholm 2021).

Consider a radiologist using AI diagnostic assistance. The system accurately flags potential abnormalities, sometimes identifying patterns the radiologist would miss. The radiologist reviews AI suggestions, applying clinical judgment to confirm or reject recommendations. When diagnoses prove accurate, who achieved the diagnostic success: the radiologist exercising supervisory judgment, the AI system whose pattern recognition enabled identification, or the collaboration irreducibly combining both?

This ambiguity affects not merely external recognition (how others credit achievement) but also self-recognition, in this sense, that workers see their capacities and effort reflected in their work. Marx identified alienation from products as estrangement from what one produces. AI collaboration creates analogous estrangement: workers may be unable to recognize completed work as distinctively *theirs* rather than outputs jointly produced with AI. This threatens the self-esteem that comes from accomplishing difficult tasks through exercised capabilities.

4.2 The Causal-Psychological Dimension: Dependency and Deskilling

The psychological concern extends beyond current achievement attribution to future capacity development. Professional expertise develops through confronting appropriately difficult challenges that stretch existing capabilities. Struggling with complex problems, making mistakes, developing judgment through experience; these processes cultivate the expertise enabling future achievement. When AI handles difficult aspects of work, these developmental opportunities contract.

This creates a dependency dynamic where workers become reliant on AI assistance for tasks they once performed independently. The worry isn't merely current achievement ambiguity but long-term capacity erosion. If AI consistently handles complex analysis, do workers retain analytical capabilities? If AI generates creative options, does human creative capacity atrophy? If AI makes difficult judgments, do workers lose the practical wisdom that emerges from exercising discretion? As in the case of coordination (Sect. 3.2), Mackenzie's relational autonomy framework illuminates what is at stake: the social scaffolding through which workers develop self-governance, including the competence conditions for authentic professional judgment, erodes when AI consistently handles precisely those tasks through which expertise develops.

Some evidence suggests AI augmentation may indeed create such dependency. Studies of GPS navigation show reduced spatial memory development in those who rely on turn-by-turn directions rather than developing internal maps (Dahmani & Bohbot 2020). Analogous concerns arise with professional AI tools: does reliance on AI-assisted research reduce lawyers' research skills? Does dependence on diagnostic AI erode radiologists' pattern recognition capabilities? Does routine use of code generation AI weaken programmers' problem-solving abilities?

The challenge is distinguishing legitimate tool use from problematic dependency. Calculators enhance mathematical work without necessarily causing deskilling; mathematicians using calculators for routine computation free cognitive resources for higher-level reasoning. Similarly, AI tools might enable professional work at higher levels of abstraction without eroding foundational capabilities. The crucial variable is whether tool use *supports* human capacity development by handling routine aspects while preserving challenging core work, or whether it *replaces* human capacity by handling precisely those difficult aspects through which expertise develops.

The distinction between tool use and problematic dependency turns on what the technology handles and what it leaves to the human user. Consider three scenarios. In the first, a tool handles peripheral aspects of a task—calculation in mathematics, spell-checking in writing—freeing cognitive resources for the core intellectual work where expertise develops. In the second, a tool handles core aspects but in ways the user fully understands and could perform independently, albeit more slowly. In this sense, a calculator is a tool used by an arithmetic-competent mathematician. In the third, a tool handles core aspects through processes the user neither understands nor could replicate—a lawyer relying on AI case analysis without understanding the legal reasoning involved.

The third scenario represents problematic dependency because it severs the connection between using a tool effectively and possessing the underlying competence the tool instantiates. The worry about AI-mediated professional work is that it increasingly resembles this third scenario: professionals use AI outputs without understanding or being able to replicate the reasoning processes involved. This matters for recognition because work-related esteem has historically been grounded in demonstrated competence, not merely in the capacity to deploy tools effectively. If the tools themselves embody the competence, the grounds for esteem shift in ways that remain culturally and institutionally unresolved.

Recent work by Denise Celentano is helpful here. She argues that automation should be designed to ensure workers can experience "contributive primary qualities"—self-direction, self-development, and social recognition—rather than optimizing solely for productivity (Celentano 2024). This means AI should serve human capacity development, handling aspects that interfere with meaningful work while preserving opportunities for the challenging engagement through which expertise grows. The goal is "fair hybrid cooperation" where human-machine collaboration enables rather than obstructs workers' development of autonomous capabilities.

4.3 The Normative-Institutional Dimension: Attribution Justice and Fair Recognition

The question of legitimate institutions concerns how to structure achievement attribution fairly in human-AI collaboration. One important resource here is work on epistemic injustice, particularly testimonial injustice, where prejudice causes deflated credibility judgments (Fricker 2007). When AI systems mediate achievement attribution, they risk encoding systematic biases that unfairly deflate certain workers' contributions while inflating others'.

Consider performance review systems that track "objective" productivity metrics. If these systems attribute collaborative achievements primarily to those whose work is easily quantified while discounting contributions that are harder to measure (facilitation, mentoring, creative problem-framing), they create systematic misrecognition. Workers whose contributions are readily trackable receive recognition while others performing equally valuable but less quantifiable work remain invisible.

This concern intensifies with AI attribution systems that pattern-match contributions to established templates. If the system was trained primarily on work styles typical of dominant groups, it may systematically undervalue contributions from those working differently. This creates what Fricker calls "hermeneutical injustice," when gaps in collective interpretive resources prevent understanding one's own experiences. Workers may lack conceptual resources to articulate how AI attribution systems disadvantage them, since the systems present as objective and technically grounded (Fricker 2007).

Addressing these institutional pathologies requires what might be called "attributional justice" (Anderson 2025a): systematic attention to whether standards involved in the achievement assessment are not themselves exclusionary or arbitrarily determined. This suggests several principles at the heart of a just form of governance of the workplace:

- *Transparency*: Workers must understand how attribution systems function and what they value. Black box algorithms making recognition decisions without explanation violate basic fairness requirements.
- *Contestability*: Workers must be able to effectively challenge attribution decisions. If AI systems determine recognition through processes workers cannot contest, this denies them standing as equals.
- *Pluralism*: Multiple forms of valuable contribution must remain visible and esteemed. Systems shouldn't encode narrow conceptions of productivity

that systematically disadvantage care work, collaborative facilitation, or innovative approaches deviating from established patterns.
- *Collective governance*: Workers should have voice in determining what gets recognized and how. Attribution systems should reflect collective judgments about valuable contribution rather than unilateral managerial definitions.

The broader institutional imperative is ensuring AI augmentation serves work's recognition-enabling functions rather than undermining them. This requires moving beyond efficiency optimization to consider whether technologies support workers' capacity to experience meaningful achievement, develop expertise through challenging engagement, and receive fair recognition for their distinctive contributions to collaborative outputs.

5 Performance Assessment: The Pathology of Synthetic Recognition

Let me begin with a note about the phrase "synthetic recognition." Several contributions to this collection also address whether AI systems themselves can be sources of recognition (see Chap. 9 by Pinsdorf 2026, Chap. 10 by Laitinen 2026 and Chap. 11 by Waelen 2026 in this volume). In particular, Rosalie Waelen's chapter, "Towards a theory of recognition from artificial others," develops a framework for what she calls artificial recognition, distinguishing genuine (attitude-based) from functional (consequence-based) forms and arguing that functionally efficacious recognitive acts by AI can carry moral weight even if ungenuine. My discussion intersects with hers at the point where AI mediates or simulates recognitive address, but it is narrower and workplace-specific. Here, I analyze the transformation of appraisal and collegial relations under algorithmic management, where there is a structural absence of reciprocity and mutual vulnerability. I therefore use the term synthetic recognition to mark these evaluative simulations in organizational settings. I prefer synthetic to artificial because it does not prejudge the acceptability of praise or feedback as "fake"; it simply signals that the recognitive address is constructed and system-generated, leaving its normative status to be argued rather than presumed.

The relationship between synthetic recognition and Waelen's artificial recognition framework warrants clarification. Waelen (Chap. 11) analyzes whether AI systems can be sources of recognition, that is, whether recognitive acts by AI can carry moral weight even if they differ structurally from human recognition. Her analysis ranges across human-AI interactions broadly conceived. My analysis is narrower: I focus specifically on workplace performance assessment where AI systems generate evaluative feedback within employment relationships characterized by power asymmetry and economic dependency.

The structural features of this context—non-reciprocity between evaluator and evaluated, consequential stakes for livelihood, absence of mutual vulnerability—make the recognitive deficits particularly acute. Waelen's functional approach may well be correct that AI-generated feedback can be "recognitively salient" in some contexts; my claim is that the specific context of algorithmic performance management exhibits pathologies that functional efficacy cannot remedy. The terms thus mark different objects of analysis rather than competing theoretical frameworks.

5.1 The Constitutive Dimension: Recognition Replaced by Evaluation

Performance assessment in traditional workplaces involved interpersonal recognition practices: supervisors acknowledged workers' contributions through feedback conversations, colleagues expressed appreciation for helpful collaboration, teams celebrated shared achievements. While these practices often occurred within hierarchical structures and served management purposes, they leave room for elements of genuine interpersonal recognition: mutual vulnerability, reciprocal acknowledgment, equality of standing between recognizer and recognized.

AI performance assessment transforms these practices fundamentally. Algorithmic systems track productivity metrics, evaluate work quality through pattern matching, assess collaboration through communication analysis, provide feedback through automated dashboards. Workers receive continuous performance information presented as objective evaluation grounded in comprehensive data analysis.

This transformation eliminates what Honneth identifies as recognition's essential features: reciprocity and mutual vulnerability (Honneth 1995; 2008a, 2008b). Algorithmic evaluation proceeds unidirectionally from invulnerable system to monitored subject. Workers are evaluated but cannot evaluate the evaluating system in turn. The algorithm assesses worker performance but is not itself vulnerable to workers' judgment in ways that would establish reciprocal recognition. This structural asymmetry means algorithmic assessment, however sophisticated, cannot provide genuine recognition; it offers what I here refer to as "synthetic recognition," the appearance of acknowledgment without the substance.

The problem isn't merely that algorithms evaluate differently than humans, but that algorithmic evaluation operates through a fundamentally different logic than interpersonal recognition. Human recognition involves vulnerable subjects acknowledging each other as equally vulnerable agents. Through this mutual vulnerability, recognition has ethical force; it represents one fallible person acknowledging another's worth. Algorithmic evaluation lacks this vulnerability structure. The system cannot be "wrong" in ways that would require reconsidering its assessment based on workers' counter-recognition. It processes data and generates evaluations without the intersubjective encounter that makes recognition recognition.

Consider the difference between a supervisor's feedback and an algorithmic performance dashboard. When a supervisor provides feedback, workers can respond, challenge assessments, provide context, appeal to shared understanding. This exchange, however asymmetric given hierarchical power, retains recognition's structure: two subjects mutually acknowledging each other even as they disagree about performance quality. An algorithmic dashboard eliminates this structure: workers receive scores and recommendations without the intersubjective space where genuine recognition occurs.

Some of the most fundamental concerns about synthetic recognition can be aptly articulated in terms of Honneth's recognition-theoretic reconceptualization of "reification," that is, treating persons as things (Honneth 2008a, 2008b). In his reconstruction of Georg Lukács's concept, reification involves "forgetfulness of recognition"—when instrumental, detached attitudes dominate while forgetting the primacy of engaged, empathic recognition that is ontologically and genetically prior. Algorithmic performance systems exemplify this reification: they treat workers as measurable data points whose performance can be assessed through quantitative analysis, forgetting the prior recognition relation that should ground any legitimate evaluation.

5.2 The Causal-Psychological Dimension: Internalization and Self-Instrumentalization

The psychological impact of synthetic recognition extends beyond current evaluation practices to affect workers' self-relation. When performance assessment becomes continuous algorithmic monitoring, workers internalize an evaluative gaze that transforms their relationship to work itself. Activity undertaken for inherent purposes (solving meaningful problems, contributing to valued outcomes, exercising expertise) becomes increasingly oriented toward optimizing metrics the algorithm tracks.

This dynamic manifests what Shoshana Zuboff calls "instrumentarian power," using machine learning to shape behavior rather than just observe it (Zuboff 2019). Algorithmic performance systems don't merely evaluate work but actively shape what workers do through feedback loops: workers adjust behavior to improve metrics, the algorithm learns from this adjustment and updates evaluation criteria, workers respond to new criteria through further behavioral modification. This creates ubiquitous control systems that undermine autonomous engagement with work.

The result is what Jaeggi identifies as self-alienation, the internal division where workers experience themselves as objects to be optimized rather than subjects pursuing meaningful purposes (Jaeggi 2014). Workers may achieve high-performance scores yet feel estranged from work itself, which has become a means to satisfying algorithmic evaluation rather than inherently purposeful activity. The psychological toll includes anxiety (constant monitoring generates performance anxiety), exhaustion (optimization pressure intensifies work), and existential emptiness (achieving metrics feels hollow compared to meaningful recognition).

This dynamic represents a distinctive failure of the causal-developmental conditions for autonomous agency identified earlier (Sect. 2.5). Where Mackenzie's framework emphasizes that self-authorization must be situated in social practices of mutual recognition, the emerging world of algorithmic performance systems substitutes optimization for recognition: workers come to understand their value not through intersubjective acknowledgment but through metrics that treat their agency as an input to be tuned rather than a capacity that they themselves cultivate.

Empirical research documents these psychological harms. Studies of gig workers subject to algorithmic management reveal severe stress from opaque rating systems affecting livelihood without transparency about evaluation criteria (Rosenblat and Stark 2016). Content moderators reviewing disturbing material under intensive productivity quotas experience significant psychological trauma, denied achievement recognition for skilled judgment while subjected to quantitative evaluation (Roberts 2019). Call center workers monitored through AI-enhanced surveillance systems report feeling reduced to "robots" whose every action is tracked and assessed.

The long-term concern is whether subjects systematically subjected to synthetic recognition retain capacity for the autonomous self-relation necessary for human flourishing. If workers internalize algorithmic evaluation logic, do they lose the ability to pursue work for inherent purposes? If performance becomes perpetually measured and optimized, does the self-respect grounding autonomous agency erode? If recognition becomes synthetic, do workers lose capacity to experience genuine interpersonal acknowledgment?

5.3 The Normative-Institutional Dimension: Legitimacy and Democratic Accountability

The institutional question concerns algorithmic performance assessment's legitimacy. In democratic societies, institutions wielding power over citizens' lives require justification grounding their authority in values those subject to power can reasonably accept. Performance assessment systems that determine workers' livelihood, reputation, and professional standing exercise substantial power. What grounds their authority?

Traditional performance assessment, despite hierarchical asymmetries, retained potential for legitimacy through procedural protections: evaluation criteria are publicly stated, workers receive explanation of assessments, appeals processes exist for contesting decisions, human judgment remains involved allowing contextual consideration. These procedures reflect basic fairness principles: those subject to evaluation should understand what is being assessed, why, and have recourse when evaluations seem unjust.

Algorithmic assessment often eliminates these legitimacy conditions. Evaluation criteria may be proprietary "black box" algorithms workers cannot examine. Assessment logic may be too complex for meaningful explanation even when disclosed. Appeals processes may involve submitting evidence to the same algorithmic system that made the contested decision, with no human judgment involved. Contextual factors the algorithm wasn't trained to consider become irrelevant to evaluation no matter how significant they seem to workers.

This opacity and unilateral control violate basic democratic principles. Workers are subjected to performance assessment systems they did not authorize, cannot understand, and cannot effectively contest—systems that profoundly affect their working lives and material welfare. This represents what Anderson calls "arbitrary power" constituting domination regardless of whether it is actively oppressive, since the capacity to wield power over others' lives without accountability violates republican freedom (Anderson 2017).

Antonio Aloisi and Valerio De Stefano's analysis of algorithmic management in platform work demonstrates how this plays out empirically: platforms use algorithmic systems to exercise extensive control while classifying workers as independent contractors avoiding employment obligations (Aloisi and De Stefano 2022). Food delivery, rideshare, and gig work involve unilateral pay determination, work assignment, and performance evaluation without worker input. Workers experience what amounts to employment relationships yet lack protections employment status provides because algorithmic mediation obscures the control being exercised.

The institutional imperative is subjecting algorithmic performance assessment to democratic governance. This includes several requirements:

- *Transparency rights*: Workers must access meaningful information about how they are being evaluated, what the system considers, how it weighs different factors. The European Union's General Data Protection Regulation (GDPR) provides a model with its "right to explanation" for automated decisions affecting individuals.
- *Human review*: Significant decisions affecting workers' lives (termination, advancement, compensation changes) should require human judgment exercising contextual discretion. Algorithms can inform these decisions but shouldn't make them unilaterally.
- *Collective information rights*: Worker representatives should access aggregate data about how algorithmic systems function, enabling collective understanding and contestation. Individual transparency is insufficient when system-level patterns create systematic disadvantages.
- *Democratic authorization*: Deployment of algorithmic assessment systems should require meaningful worker participation in decisions about whether to implement them, how to design them, what purposes they serve. This means moving from unilateral management prerogative to collective governance of performance evaluation practices.

The broader institutional principle is that performance assessment should function as genuine recognition—interpersonal acknowledgment of valued contribution—rather than unilateral algorithmic evaluation. This doesn't prohibit using AI to support assessment (providing data for human evaluators, identifying patterns for consideration, enabling more comprehensive review). But it requires preserving the intersubjective structure that makes recognition recognition: reciprocal acknowledgment between vulnerable subjects who see each other as equals despite hierarchical differences in workplace authority.

6 Conclusion: Recognition, Alienation, and the Future of AI-Mediated Work

6.1 Recognition Theory's Distinctive Contribution

This chapter has sought to highlight recognition theory's relevance for illuminating what makes AI-mediated labor troubling even when productivity increases. Conventional frameworks focus on quantifiable concerns: job displacement, wage effects, skill requirements. While these material issues matter profoundly, they miss workers' articulated experiences: the strangeness of receiving "recognition" from algorithms, the anxiety of perpetual evaluation, the difficulty of claiming achievements in human-AI collaboration, the loss of collegial bonds through algorithmic coordination.

Recognition theory explains why these experiential dimensions are not merely subjective dissatisfaction but indicate real normative problems. Work matters not only for income but for identity formation, social integration, and democratic cultivation (Waelen 2025a). Through work, individuals gain self-esteem through acknowledged contribution, develop capacities for autonomous agency, and learn to recognize others as both vulnerable and capable, which Honneth has analyzed as the essence of social freedom (Honneth 2014, 2024a, 2024b). When AI transformation threatens these functions, something more than economic welfare is at stake: the social conditions supporting both personal flourishing and democratic capacity. The added value of this analysis is that recognition theory identifies who owes what to whom in workplace relations (it generates obligations, not just diagnoses), whereas alienation theory describes structural pathologies without generating the same specificity about institutional remedies. Recognition theory yields the design principles; alienation theory describes the conditions that make them necessary.

The analysis across three dimensions—coordination/collaboration, quality/achievement, and performance assessment—reveals AI's recognition impacts operate at multiple levels. Constitutively, AI changes what work *is* by transforming collaborative processes into algorithmic coordination, blurring achievement boundaries in human-AI collaboration, and replacing interpersonal recognition with algorithmic feedback or "synthetic recognition." Causally-psychologically, these transformations erode workers' relational autonomy, create dependency and potential deskilling, and generate self-alienation through internalized optimization pressures. Normatively-institutionally, AI deployment without democratic authorization violates basic fairness principles while threatening the workplace democracy Honneth identifies as constitutive for political democracy (Honneth 2023, 2024a, 2024b).

Of course, in focusing here on the workplace-level analysis of the dynamics of injustice, I have bracketed out structural, societal, and global forces that shape these dynamics fundamentally. I do not wish to deny that there are many structural conditions in the background that are also legitimate and even urgent targets for critique. Jaeggi's alienation framework, for example, provides a good example of an analysis that complements the level of analysis pursued here by capturing structural features recognition theory alone cannot address (Jaeggi 2014, 2023). I have focused here on the way in which recognition theory analyzes primarily interpersonal and institutional misrecognition, how AI systems deny workers genuine acknowledgment. For a complete analysis, attention must clearly be paid to wider forces, also of the nature of capitalism as the context within which AI technologies are deployed (Fraser and Jaeggi 2018).

6.2 Opportunities and Concern

Assessment of these three transformations requires differentiation rather than wholesale condemnation or celebration. I conclude here by highlighting a few key points.

Performance assessment may well present a particularly important concern from this perspective Synthetic recognition represents a distinctive pathology difficult to address through better design. The structure of algorithmic evaluation—unidirectional feedback from invulnerable system to monitored subject—seems fundamentally incompatible with genuine recognition's reciprocity. Mitigation requires preserving human evaluators as primary sources of recognition, using AI only as support tools that enhance rather than replace human judgment. The goal should be augmenting human capacity for contextual, reciprocal recognition rather than automating assessment entirely. As Honneth emphasizes, recognition requires mutual vulnerability and reciprocal acknowledgment between subjects who see each other as equals (Honneth 1995, 2008a, 2008b), conditions algorithmic systems cannot satisfy. That said, Waelen's functionalist proposal (Waelen, Chap. 11) helps explain why people sometimes *experience* system-generated feedback as morally relevant even when it falls short intersubjectively. To keep terms clear in what follows, I reserve *synthetic recognition* for these managerial-evaluation contexts marked by non-reciprocity, while *artificial recognition* names the broader family of human–AI interactions whose functional effects Waelen theorizes.

Although discussions of the transformation of the understanding of achievement are mor prominent, the concern is more ambiguous While blurred boundaries create genuine problems for professional identity and self-esteem, human-AI collaboration may eventually develop new forms of professional identity based on effective partnership rather than autonomous mastery. This requires cultural transformation, that is, reconceiving expertise as skill in collaborative orchestration rather than individual achievement. Institutional support is needed: fair attribution practices acknowledging human contribution within collaboration, equitable access to tools so recognition patterns don't advantage only those with premium AI access, explicit recognition of collaboration competencies as valuable professional skills. Celentano's "fair hybrid cooperation" framework offers guidance: automation should enable workers to experience contributive primary qualities—self-direction, self-development, dignity, and recognition (Celentano 2024).

In many ways, AI-mediated improvements to coordination seem to offer the most opportunity for positive transformation, though this requires deliberate design choices. AI tools that facilitate rather than replace collaboration—enabling coordination across distance while preserving space for genuine interaction—could expand rather than contract opportunities for mutual acknowledgment. The key is preserving what Jürgen Habermas called "communicative action" (interaction oriented toward mutual understanding) while using AI to handle routine coordination tasks. For instance, AI might automate scheduling logistics while flagging issues requiring collaborative deliberation, or provide information synthesis while requiring human teams to discuss implications and determine responses. The goal is supporting the organic solidarity Durkheim identified: helping workers understand their interdependence and develop reciprocal bonds while avoiding the fragmentation that destroys cooperative experience.

6.3 Design Principles and Institutional Imperatives

Several principles emerge for recognition-sensitive AI integration, though articulating them fully would require a separate treatment. Here is a first sketch:

Democratic authorization: Workers must have meaningful voice in AI deployment decisions affecting their work. This isn't merely consultation but genuine participation in determining which technologies are appropriate, how they're implemented, what values they serve. Arbitrary power over workers' lives violates republican freedom regardless of whether it is actively oppressive. Democratic control is not merely desirable but essential for legitimate technological governance.

Recognition preservation: Systems should enhance rather than displace human acknowledgment. AI can support recognition by making contributions more visible, facilitating cross-team awareness, enabling better-informed appreciation, but shouldn't replace the intersubjective process through which colleagues recognize each other. The goal is preserving what O'Neill and Smith characterize as social freedom, which for them means freedom actualized through recognized participation in cooperative relationships (O'Neill and Smith 2022).

- *Pluralism support*: Multiple forms of valuable contribution must remain visible and esteemed. Systems shouldn't encode narrow conceptions of productivity that systematically disadvantage care work, collaborative facilitation, or innovative approaches deviating from established patterns. This addresses Fricker's concern about hermeneutical injustice—ensuring workers have conceptual resources to articulate their contributions even when they don't fit dominant templates (Fricker 2007).
- *Transparency and contestability*: Workers must understand how AI systems function and be able to effectively challenge decisions. Black box algorithms wielding power over work lives violate basic fairness requirements. As Aloisi and De Stefano argue, transparency rights and human review of automated decisions are essential for restoring worker agency (Aloisi and De Stefano 2022).
- *Social integration support*: Technologies should sustain rather than fragment workplace communities. The goal isn't merely coordinating isolated individuals but supporting the relational context through which work functions as a sphere of mutual recognition and social freedom. This means preserving opportunities for the collaborative engagement that generates organic solidarity.

These aren't merely technical design guidelines but political demands requiring transformed power relations. As with historical struggles for the 8-h workday or workplace safety standards, achieving recognition-sensitive AI integration requires social movements, regulatory frameworks, and cultural shifts in how we understand technology's proper role in work.

To illustrate how these principles might translate into concrete institutional arrangements, consider two specific proposals.

First, algorithmic impact assessments for workplace AI should include recognition dimensions alongside efficiency metrics. Before deploying coordination, attribution, or evaluation systems, organizations should assess: Does this system preserve opportunities for workers to understand their contribution to collective outputs? Does it maintain space for interpersonal acknowledgment? Does it allow meaningful contestation of assessments? Such assessments, modeled on environmental or privacy impact assessments, would create institutional pressure to consider recognition effects that market pressures alone will not generate.

Second, recognition preservation requirements could mandate that certain decisions remain subject to human judgment regardless of algorithmic recommendations. Just as financial regulations require human approval for certain transactions, workplace regulations could require that termination decisions, significant performance evaluations, and major task reassignments involve human evaluators capable of contextual judgment and responsive to worker contestation. This does not prohibit algorithmic inputs but preserves the intersubjective structure that genuine recognition requires.

These proposals are illustrative rather than exhaustive, but they suggest that recognition-sensitive AI governance is institutionally feasible, not merely philosophically desirable.

6.4 Critical Theory's Contribution

Recognition theory reveals that work matters not merely for material provision but for identity formation, social integration, and democratic cultivation. Through work, individuals gain self-esteem through acknowledged contribution, develop capacities for autonomous agency, and learn to recognize others as both vulnerable and capable—the essence of social freedom. When AI transformation threatens these functions, something more than economic welfare is at stake: the social conditions supporting both personal flourishing and democratic capacity.

This is why recognition-theoretic analysis contributes distinctively to understanding AI's workplace impact. Efficiency-focused frameworks can calculate productivity gains and job displacement but cannot illuminate what workers themselves articulate (Waelen 2025b): the strangeness of synthetic recognition, the anxiety of perpetual algorithmic evaluation, the difficulty of claiming achievements in hybrid human-AI collaboration, the loss of collegial bonds through algorithmic coordination. These experiential dimensions matter because work's normative significance extends beyond income to encompass identity, dignity, and the cultivation of social freedom.

The struggle for recognition in AI-mediated workplaces thus becomes a key site for contemporary critical theory and is continuous with Frankfurt School traditions of analyzing technology's role in domination and emancipation while addressing novel configurations of power and possibility (Waelen 2022). It requires sustained theoretical analysis of how algorithmic systems affect the practical relations-to-self constitutive of human agency. It demands attention to subtle pathologies emerging when efficiency optimization displaces recognition, when coordination replaces collaboration, when synthetic acknowledgment replaces genuine intersubjective encounter. And it calls for practical efforts to democratize workplace AI's design and deployment, ensuring technological transformation serves human emancipation rather than introducing new forms of domination concealed beneath productivity gains. As Jean-Philippe Deranty has convincingly argued (against post-work theorists who see automation as enabling work's obsolescence), we should advocate for democratic control over technology to preserve work's recognition-enabling functions (Deranty 2024, Boston 2018). This chapter is very much in line with that aspiration: the goal isn't eliminating work through AI automation but democratically governing AI to ensure work retains its essential role as a sphere of mutual recognition supporting both personal autonomy and democratic citizenship.

The path forward requires moving beyond both technological determinism (AI necessarily destroys recognition) and naive optimism (AI is a neutral tool for efficiency). Instead, recognition depends on democratic governance structures that subject AI deployment to collective control, preserve space for genuine interpersonal acknowledgment, and ensure technological development serves human flourishing rather than treating humans as means to productivity optimization. This represents not merely a technical challenge but a political struggle continuous with labor movements' historical efforts to secure dignity, recognition, and democratic voice in the workplace.

References

Aloisi, Antonio, and Valerio De Stefano. 2022. *Your Boss Is an Algorithm: Artificial Intelligence, Platform Work and Labour*. Oxford: Hart Publishing.

Anderson, Elizabeth. 1999. What Is the Point of Equality? *Ethics* 109 (2): 287–337.

Anderson, Elizabeth. 2017. *Private Government: How Employers Rule Our Lives (and Why We Don't Talk About It)*. Princeton: Princeton University Press.

Anderson, Joel. 2024. Attributional Justice, Resilient Resistance, and Recognition Supports: Extending Catriona Mackenzie's Three-Dimensional Account of Autonomy for Critical Social Theory. In *Paper Presented as the 2024 Annual Meeting the Dutch National Research School (August 31, 2024)*.

Anderson, Joel. 2025a. Autonomous Agency, Social Justice, and Interpersonal Recognition. In *Ethics of Behaviour Change Technologies*, edited by Joel Anderson, Lily Frank, and Andreas Spahn. Bloomsbury Academic.

Anderson, Joel. 2025b. Erkenning en vervreemding op de digitale werkplek: AI-gemedieerde arbeid. *Wijsgerig Perspectief* 65 (2): 15–23.

Boston, Timothy. 2018. New Directions for a Critical Theory of Work: Reading Honneth Through Deranty. *Critical Horizons* 19 (2): 111–124.

Celentano, Denise. 2024. Labor Automation for Fair Cooperation: Why and How Machines Should Provide Meaningful Work for All. *Journal of Applied Philosophy* 41 (2): 309–328.

Celikates, Robin, Axel Honneth, and Rahel Jaeggi. 2023. The Working Sovereign: A Conversation with Axel Honneth. *Journal of Classical Sociology* 23 (3): 318–338.

Christiaens, Tim. 2022. *Digital Working Lives: Worker Autonomy and the Gig Economy*. Rowman & Littlefield.

Dahmani, Louisa, and Véronique D. Bohbot. 2020. Habitual Use of GPS Negatively Impacts Spatial Memory During Self-Guided Navigation. *Scientific Reports* 10:6310.

Danaher, John, and Sven Nyholm. 2021. Automation and the Achievement Gap: Resolving a Paradox in the Ethics of Automation. *AI and Society* 36:1005–1019.

Deranty, Jean-Philippe. 2024. *The Case for Work*. Cambridge: Polity Press.

Durkheim, Émile. 1984. *The Division of Labor in Society*. Trans. W.D. Halls. New York: Free Press.

Fraser, Nancy, and Rahel Jaeggi. 2018. *Capitalism: A Conversation*. John Wiley & Sons.

Frega, Roberto, Lisa Herzog, and Christian Neuhäuser. 2019. Workplace Democracy – The Recent Debate. *Philosophy Compass* 14 (4): e12574.

Fricker, Miranda. 2007. *Epistemic Injustice: Power and the Ethics of Knowing*. Oxford: Oxford University Press.

Hirvonen, Onni, and Keith Breen. 2020. The Wages of Sin: Recognition, Meaningful Work, and Decent Pay. *Res Publica* 26:277–295.

Honneth, Axel. 1995. *The Struggle for Recognition: The Moral Grammar of Social Conflicts*. Trans. Joel Anderson. Cambridge, MA: MIT Press.

Honneth, Axel. 2008a. Arbeit und Anerkennung: Versuch einer Neubestimmung. *Deutsche Zeitschrift für Philosophie* 56 (3): 327–342.

Honneth, Axel. 2008b. *Reification: A New Look at an Old Idea*. Oxford: Oxford University Press.

Honneth, Axel. 2014. *Freedom's Right: The Social Foundations of Democratic Life*. Trans. Joseph Ganahl. New York: Columbia University Press.

Honneth, Axel. 2023. Democracy and Fair Labor Conditions. *Journal of Business Ethics* 184 (1): 1–10.
Honneth, Axel. 2024a. *The Working Sovereign: Labour and Democratic Citizenship*. Trans. Daniel Steuer. Cambridge: Polity Press.
Honneth, Axel. 2024b. A Genealogy of Work: Review Essay on Raymond Geuss's "Über die Arbeit. Ein Essay". *Filozofia* 79 (1): 5–17.
Hopster, J. K. G. 2024. Socially Disruptive Technologies and Epistemic Injustice. *Ethics and Information Technology* 26 (1): 14.
Jaeggi, Rahel. 2014. *Alienation*. Trans. Frederick Neuhouser and Alan E. Smith. New York: Columbia University Press.
Jaeggi, Rahel. 2023. Sharing the Universal Resources: Remarks on Honneth's Understanding of Work. *Journal of Classical Sociology* 23 (2): 195–208.
Laitinen, Arto. 2026. Mutual Recognition Between Humans and Robots? Common Sense Phenomenology Versus the Scientific Worldview. In *Normative Relations of Recognition in the Age of AI*, edited by Rosalie Waelen, 215–242. Springer.
Löhr, Guido. 2023. Conceptual Disruption and 21st Century Technologies: A Framework. *Technology in Society* 74:102327.
Mackenzie, Catriona. 2014. Three Dimensions of Autonomy: A Relational Analysis. In *Autonomy, Oppression, and Gender*, 15–41. Oxford University Press.
Mackenzie, Catriona, and Natalie Stoljar, eds. 2000. *Relational Autonomy: Feminist Perspectives on Autonomy, Agency, and the Social Self*. New York: Oxford University Press.
Marx, K. 1988. Economic and Philosophic Manuscripts (1844) (L. D. Easton & K. H. Guddat, Trans.). In *Marx Selections*, edited by A. W. Wood. Macmillan.
Marx, Karl. 2007. *Economic and Philosophic Manuscripts of 1844*. Mineola, N.Y: Dover Publications.
O'Neill, Shane, and Nicholas H. Smith. 2022. Social Freedom as the Purpose of the Modern University. *Philosophy and Theory in Higher Education* 4 (1): 1–23.
Pinsdorf, Christina. 2026. Moral Recognition Relations Between Lifeforms and AI?. In *Normative Relations of Recognition in the Age of AI*, edited by Rosalie Waelen, 187–213. Springer.
Roberts, Sarah T. 2019. *Behind the Screen: Content Moderation in the Shadows of Social Media*. New Haven: Yale University Press.
Rosenblat, Alex, and Luke Stark. 2016. Algorithmic Labor and Information Asymmetries: A Case Study of Uber's Drivers. *International Journal of Communication* 10:3758–3784.
Van de Poel, Ibo, Dominic Lenzi, Jeroen Hopster, Behnam Taebi, Julia Hermann, Elena Ziliotti, Lily Frank, and Sven Nyholm, eds. 2023. *Ethics of Socially Disruptive Technologies: An Introduction*. Cambridge, UK: Open Book Publishers.
Viehoff, Juri. 2025. Solidariteit en de gedigitaliseerde werkplek. *Wijsgerig Perspectief* 65 (2): 15–23.

Waelen, Rosalie. 2022. Why AI Ethics is a Critical Theory. *Philosophy and Technology* 35 (1): 20.
Waelen, Rosalie. 2025a. Rethinking Automation and the Future of Work with Hannah Arendt. *Journal of Business Ethics*.
Waelen, Rosalie. 2025b. The Desirability of Automizing Labor: An Overview. *Philosophy Compass* 20.
Waelen, Rosalie. 2026. Towards a Theory of Recognition from Artificial Others. In *Normative Relations of Recognition in the Age of AI*, edited by Rosalie Waelen, 243–258. Springer.
Zuboff, Shoshana. 2019. *The Age of Surveillance Capitalism: The Fight for a Human Future at the New Frontier of Power*. New York: PublicAffairs.

AI, Recognition, and Identity: The Normative and Epistemic Dimensions of Identity Formation Through Algorithms

Natalia Juchniewicz

Abstract The aim of this chapter is to answer the question of whether AI can be a carrier of recognition and, if so, whether it can help co-shape individual and collective identity. To answer these questions, I distinguish between recognition in the normative (Honneth, Taylor) and epistemic meaning (Butler) and raise the issue of normative and epistemic power of AI. By introducing Butler's concept of recognizability and interpellative power, I show that AI forms the algorithmic discourse in which individual and collective identity is presented. This undoubtedly has normative consequences, but it may also provide a basis for treating AI as a hermeneutic tool through which we can see ourselves in non-obvious categorisations and identifications. AI recognition of a specific user or group should not be considered solely in normative terms, but also as a form of epistemic representation of our identity. We decide whether our data doubles are a form of (mis)recognition and/or an impetus to reflect on who we are. As I show in the chapter, it is unrealistic to expect AI to fully recognise us in all aspects of our identity, as we never receive full recognition in social reality either. However, it is an important technology for the achievement of partial or distributed recognition in the normative and epistemic meaning.

N. Juchniewicz (✉)
Faculty of Philosophy, University of Warsaw, Warsaw, Poland
e-mail: n.juchniewicz@uw.edu.pl

R. Waelen (ed.), *Normative Relations of Recognition in the Age of AI*, Philosophy and Ethics of AI 2, https://doi.org/10.1007/978-3-032-34028-3_4

Keywords AI (artificial intelligence) • Algorithms • Normative recognition • Epistemic recognition • Identity formation • Interpellative power • Epistemic power • Hermeneutic tool

1 Introduction

In the theories of recognition it is assumed that a human being builds self-understanding through intersubjective relationships with other people. It is impossible to know who a person is without other subjects with whom different meanings of identity, values and virtues can be shared. It is, however, a matter of concern among scholars focused on the concept of recognition what exactly the carrier of recognition is. In this article, I raise the question of whether artificial intelligence and learning algorithms can be a carrier of recognition and, therefore, whether it is possible to co-shape subjective and collective identity through them. There are two problems associated with this question: the first problem concerns the meaning of the recognition exhibited by artificial intelligence (AI) when interacting with a person. Namely, are we confronted with a purely normative meaning of recognition by AI or an epistemic meaning? The second problem concerns whether AI recognition or misrecognition may be a catalyst for reflection on one's own identity and, therefore, whether data doubles generated by AI constitute a new hermeneutic tool.

To answer the central research question and address its associated problems, the chapter is structured as follows. In Sect. 2, I will present how the classical theory of recognition represented by Axel Honneth, Charles Taylor, and Judith Butler defines the significance of recognition for the formation of individual and group identity and I will show the difference between the normative and epistemic meanings of recognition. Drawing from Butler's theory, I will demonstrate the distinction between recognition and 'recognizability' and explain why artificial intelligence can be considered a new form of interpellative power that co-shapes our identity. In Sect. 3, I will focus on presenting how the normative consequences of AI are analysed in the literature on the subject, which refers to recognition theory. My aim will be to highlight the normative and epistemic power of algorithms. AI allows for the identification and categorisation of data about specific individuals and groups, which involves the reproduction of biases and stereotypes, and even influences the conditions for the reproduction of social justice. As I will demonstrate, AI algorithms not only influence recognition in a normative meaning but also constitute a new, discursive power which reveals the epistemic dimension of recognition. In Sect. 4, I will present that the AI's algorithms influencing the conditions of identity formation can and should be examined as a hermeneutic tool. This is because artificial intelligence refers not so much to us as to our data generalised quantitatively in relation to other data. Our datafied identity is therefore a digital representation of our data, which has two consequences. On the one hand, the categorisations made on data by AI have real normative consequences, but on the other hand, it is up to users to decide whether their data double constructed by AI will be treated as an expression of recognition or not. In Sect. 5, I draw attention to the fact that the essence of recognition relation is reciprocity and reflexivity. A conclusion that can be drawn from a reflexive analysis of recognition is that full recognition in social life, let alone through technology, is unattainable. The point is to constantly engage in a struggle to become a subject in ever newer social, political and technological conditions. In Sect. 6, I summarise the most important arguments of the chapter.

2 Normative and Epistemic Meaning of Recognition

The core idea of recognition theories is that recognition from others is essential for individual persons or groups to build identity. Self-understanding and hence self-identity can be shaped only through the recursive relation with other people. A human being, to know who he/she/they is has to open towards others but also has to come back to itself on the higher, reflexive level. This conception of recognition, inspired by G.W.F. Hegel and earlier by J.G. Fichte, has been formulated in the most extensive manner by Axel Honneth and Charles Taylor and developed by many interpreters of these theories; (Hirvonen 2017; Ikäheimo 2002; Ikäheimo and Laitinen 2011; Laitinen 2002). In the original version described by Hegel, the idea of recognition is based on the assumption that no one could know themselves without being recognised as a subject by another subject (Hegel 1977, 1983). In other words, we are not born as subjects, we become subjects in the process of building the self through the interactions and relations with others. This emphasises the fact that a subject is not a given, ontological fact but an effect of phenomenological, social and political process of self-creation. The Hegelian idea of the spirit (*Geist*) is a result of such thinking about the subject as a project and not as an established entity (Kislev 2020). This theory had a great influence on social theory, because it gave a possibility to look at a subject as socialized by different institutions and other people which take part in the process of this self-development of a person.

The abovementioned Hegelian perspective on human being has been extended in the theory of Axel Honneth, who emphasised that the procedural character of a subject's development is expressed through three types of relations-to-self, which manifest in three, different modes of recognition. The first of these is self-confidence, which develops through emotional support in family relationships. Family guarantees to the individual, a fundamental element for building self-certainty, which is love. It not only involves recognising the value of another person as a member of our own, most intimate community but also the practices of care that allow us to understand that we can, on the one hand, expect help from others in satisfying our fundamental needs, and on the other hand, learn to reciprocate that assistance (Honneth 1995, 95–96). The second type of relation-to-self is self-respect, which we achieve through the recognition by others of our cognitive abilities that translate into our capacity to make decisions about ourselves. This, as d by Honneth, holds fundamental significance for understanding the rights as an expression of individual freedom and the expectation of individuals to be recognised by law in a suitable manner. Here too, as a result of legal recognition, reciprocity is established, determining the rights of the individual as well as their obligations towards others (Honneth 1995, 108). The third type—self-esteem—assumes that, in addition to the subjectivity of being recognised within the family and the generality of respect we achieve through the law, there is also a particular form of the struggle for recognition that is connected to our sense of uniqueness and social usefulness. Honneth observes that in the context of the struggle in economic, social and political conditions, solidarity is essential for symmetrically respecting the fact that we all participate in each other's life plans, aspirations, and hopes. The competition in the job market does not have to lead to treating other people as adversaries, but rather, it can serve as the basis for mutual understanding that within certain professional groups, we face similar challenges and follow similar paths of development, which fosters our mutual understanding (Honneth 1995, 128).

Charles Taylor emphasises that recognition forms the basis for building the identity of both individuals and collective entities. According to him, a key achievement of the Enlightenment is the realization that a subject is free in defining themselves as long as the chosen way of life does not limit the freedom of others.

> Before the late eighteenth century, no one thought that the differences between human beings had this kind of moral significance. There is a certain way of

> being human that is my way. I am called upon to live my life in this way, and not in imitation of anyone else's life. But this notion gives a new importance to being true to myself (Taylor 1994, 30).

Living in accordance with oneself is what Taylor calls authenticity, which in practical terms can mean both the desire to build one's identity through unification with others (at various levels of group identification) and through differentiation from others (also at various levels of group identification). This leads to two different recognition-based policies: the first is based on understanding universal human potential (Taylor 1994, 41–42), which is the foundation of modern human dignity theory, as 'difference-blind'. Rights, norms, and principles should be equal for all because, in principle, human potential and the directions of its development are equal. The second, on the contrary, emphasises differences to underscore that ways of life and self-definition should respect the differences among people. As Taylor demonstrates, even though both types of policies are rooted in principles of respect for the subject, they define it in entirely different ways in practice. Taylor, using the concept of the general will by Rousseau and Hegel's theory of recognition, suggests that the solution to this conflict should involve realizing that 'there is a "we" that is an "I", and an "I" that is a "We"' (Taylor 1994, 50). Therefore, there is no individual without a broader cultural and social reference, and there are no societies without individuals practicing them. Mutual recognition should not only involve respecting individual persons but also valuing entire cultures and treating not only individuals equally but also entire communities.

The theories of recognition in both of the abovementioned interpretations highlight predominantly the normative consequences of being or not being recognised. Therefore, these two interpretations had a great contribution to the social studies of struggles and protests, colonial, gender and racist issues, individual and group identities, etc. (Badarin 2023; Balaton-Chrimes and Stead 2017; Chari 2004; Mcintosh 2022; Fricker 2018; Fraser and Honneth 2003). However, as it is emphasised by Judith Butler in her correspondence with Honneth (Butler 2021b), to talk about normative consequences we have to be treated as subjects who are given a freedom of expression and who are treated equally by the rest of the society. Thus, as Butler argues, the theory of recognition in its normative approach does not explain what the source of these norms is. She emphasises the difference between recognition and 'recognizability'—if we assume that recognition is normative in nature, then recognizability means the epistemic basis for being included in a normative community in the first place by being identified as a human being, as able to speak, as equal etc. '(...) what are the existing norms that establish recognizability, that is, the field in which recognition proves to be possible?' (Butler 2021b, 43). This difference is of great importance for the application of recognition theory to AI research, as I will show below, because the recognition that AI grants to humans is much more consistent with the meaning of recognition proposed by Butler than with that proposed by Honneth or Taylor.

The problem of the meaning of the concept of recognition is probably the most discussed issue in the literature on recognition. Honneth states that the assumption that recognition (*Anerkennung*) could be understood as recognising (*erkennen*) is not a Hegelian approach and is not the aim of Honneth's own perspective: 'the mere act of ascribing a certain property or characteristic to someone isn't sufficient to call it an occurrence of recognition in the Hegelian sense (...)' (Honneth 2021, 24). He even states that such recognition through identification has no normative consequences (Honneth 2021, 23). Honneth tries to be in accordance with the Hegelian tradition, to focus primarily on the social norms as shaping our self and building a sensitive net of social relations which can be a source of both recognition and misrecognition. This intention might be summed up as the research from the 'inside' of the society and less from the 'outside'—so not from the conditions which frame the concrete form of the society, but from the perspective of existing cultural, social and political differences (assuming, of course, that norms and values of such society might change in time). Butler, on the other hand, focuses on the 'outside' or, what is called, 'the frames of recognition' (Ikäheimo et al. 2021, 11). Honneth also emphasises an important linguistic difference between the English and German understanding of the concept of recognition. The German meaning of recognition boils down to granting someone a certain status or right, while the English meaning of the term refers to an epistemic act of reidentification, which can only secondarily have a normative character (Honneth 2021, 25–26). This difference, in my opinion, emphasises that Butler's epistemically oriented theory fits much more broadly into the issue of recognition in the context of AI than Honneth's or Taylor's theories alone, as it allows us to raise the question of how AI operates as a specific technology in relation to the subject, rather than merely defining the normative consequences of this relationship.

In her theory, Judith Butler emphasises the important role of recognizability as a source of epistemic problems and epistemic understanding of recognition.

> One comes to <<exist>> by virtue of this fundamental dependency on the address of the Other. One <<exists>> not only by virtue of being recognized,

> but, in a prior sense, by being recognizable. The terms that facilitate recognition are themselves conventional, the effects and instruments of a social ritual that decide, often through exclusion and violence, the linguistic conditions of survival subjects (Butler 2021a).

Butler, what is important from the perspective of AI, asks the question about the carrier of recognition and indicates that the conditions of recognition are prior to the act of recognition and are expressed through language. Language has a performative power since people are embedded in the values, discourse and practices which are rooted in the language (Butler 2021a, 8). 'We belong to each other prior to the act of recognition that constitutes our value for each other. We do not make up those terms through which we are recognized since we are born into a language that precedes us and acts on us' (Butler 2021b, 33).

The power of language and performative speech can be explained through the reference to Althusser (Butler 2021a, 24–39). In the text *Ideology and Ideological State Apparatuses* Althusser analyses what ideology is and how it produces the subject. He states that 'all ideology hails or interpellates concrete individuals as concrete subjects, by the functioning of the category of the subject' (Althusser 2001, 162). The interpellation, as Althusser explains, transforms individuals into subjects when it identifies individuals with the concrete categories of people. A famous example of the interpellation of a policeman hailing 'Hey, you there' and the fact that usually a person hailed will turn round explains, according to Althusser, that in the moment of turning round an individual becomes a subject. 'Because he has recognized that the hail was "really" addressed to him, and that "it was really him who was hailed" (and not someone else)' (Althusser 2001, 163). The fact, however, that policeman hailed was the effect of epistemic recognition of a person with a possible criminal or someone suspected which is often built on existing social biases. 'The act of recognition becomes an act of constitution: the address animates the subject into existence' (Butler 2021a, 26) even if this recognition must not always be correct. Butler continues the analysis of interpellations and the power of language which is given to different institutions and social roles indicating that performative speech constitutes the subject on a very basic level of human identity.

> The doctor who receives the child and pronounces—'It's a girl'—begins that log string of interpellations by which the girl is transitively girled: gender is ritualistically repeated, whereby the repetition occasions both the risk of failure and the congealed effect of sedimentation (Butler 2021a, 49).

According to Butler, gender identity is shaped by identification with a gender, which is *de facto* determined by a doctor or the child's parents, and is then performed in social practices, language, cultural codes, etc. The same process might be observed with racial, ethnic or religious biases which are the effect of interpellations of power in the hands of priests, politicians, judges, teachers or journalists etc. even if the interpellative power itself does not come directly from these persons, but they are merely its representatives ('The power (...) precedes the "one" who speaks such power, and yet the one who speaks nevertheless appears to have that power' (Butler 2021a, 49)).

If we look at AI and how it works, we could see that this technology fulfils the criteria of power which uses performative language of interpellations analysed by Butler and Althusser. The moment I open any app or website I am subjected to the terms of the algorithms which influence me but are not created by me. Given that all users are subjected in the same way, the rules of algorithmic AI are beyond our control. The power of AI is everywhere but is invisible, similar to the rules which govern any human language. Moreover, even if AI does not literally hail on me 'hey you there', the digital content which I have access to is a form of such hailing. It recognises me as a subject with different interests, thoughts or (dis)likes, but I also 'turn round' towards this content when I simply dive into it, accept it and follow the suggestions of AI.

We 'belong' to 'terms' and 'language' of AI that 'precedes us and acts on us' and this way AI frames the conditions of recognition we could achieve in a technologized society as users of technology. In the next section of the chapter, I will show what normative and epistemic consequences this has for the understanding of recognition and for shaping human identity in the context of AI use.

3 AI as a Normative and Epistemic Power

As Rosalie Waelen has shown, artificial intelligence primarily identifies and categorises data, which leads, however, to specific normative consequences. Using the example of facial recognition technology, Waelen shows how misidentification and miscategorisation lead to misrecognition. Biometric facial analysis technologies are significantly more likely to misidentify people who are darker-skinned, female, and young (Waelen 2022, 5) as 'suspicious', leading to serious abuses of rights and civil liberties. Miscategorisation of people, such as assigning Asian facial features to the 'blinking' category, undermines respect. As Waelen notes: 'These are vivid examples of how miscategorization is not just an inconvenient or innocent error, but truly demeaning and potentially damaging for a person's self-esteem' (Waelen 2022, 5). Waelen also identifies a third dimension of misrecognition that is not strictly related to the technical aspect of AI technology, but rather to the reduction of an individual's identity to specific qualities analysed by the technology. Human identity can be subject to misrecognition when technology makes technical errors, but in some cases misrecognition also consists in being completely transparent to technology. When AI fails to take into account the characteristics and qualities of specific individuals or groups, when it is unable to adequately represent their identity in the way it operates, misrecognition also occurs.

> So, in short, facial recognition is limited when it comes to recognizing people's subjective identity, as the technology can only recognize objective forms of identity and those aspects that can be read of a person's face. Facial recognition is often promoted and used as a tool for personalization, but given its limitations when it comes to recognizing a person's true individuality or individual uniqueness, we should question the extent to which facial recognition can really personalize services (Waelen 2022, 6).

Similarly, Waelen and Wieczorek point to three types of gender biases in AI systems: literal misrecognition (of, for example, faces or voices); reproduction of stereotypes; and exclusion of the needs, values, and perspectives of specific user groups (Waelen and Wieczorek 2022, 53). These problems are directly related to the misrecognition of women by AI technology. However, solving the problem of being inadequately recognised by AI leads the authors to two fundamental aspects of this issue: the design of the technology (who trains AI, how, and on what data) and broader social issues (struggling with gender stereotypes in social life) (Waelen and Wieczorek 2022, 52–53). The research conducted by Waelen and Wieczorek shows that the concept of recognition can be approached in the context of AI either epistemically, in which case the solution to the problem of misrecognition should be sought in the appropriate design of the technology, or normatively, where we expect technology to respect, for example, the principles of equality and social justice in its operation. However, as I will show below, AI technology itself is already part of a certain power order, a form of shaping a specific discourse, an algorithmic representation of social reality, which is why simply drawing attention to the normative consequences of AI is not enough to understand the importance of this technology for shaping the identity of individuals and groups. It is necessary to point out its epistemic power too.

Butler emphasises that the interpellative power of language is insensitive to truth and falsehood; it is not interested in reproducing reality, but only in establishing a specific order and, consequently, a specific type of subject—a subject subordinate to power itself. Importantly, in the context of AI, interpellative power does not have to be exercised by a specific person, as it is not the person who is the source of power, but the discourse. This discourse, however, is crucial for the formation of the subject. '(...) discursive power (...) operates without a subject, but that constitutes the subject in the course of its operation' (Butler 2021a, 35). AI algorithms currently have similar power over users of this technology—they introduce an algorithmic discourse; rules that users must submit to if they want to use a particular technology. Rafanelli (2022), although she does not refer to Butler, accurately diagnoses the need to question the power that algorithms have over us.

Rafanelli focuses on the relation between AI and the problem of justice and power. She diagnoses, first of all, discrimination and structural injustice as a problem which emerges from the AI technologies which are trained on structurally biased data. The fact that, for instance, AI which is trained on resumes submitted to Amazon and prioritizes men over women, is the effect of the quality of data delivered to it which was predominantly based on men's resumes (Rafanelli 2022: 3). One could state that this is only the problem of data, but the real, social and economic consequences of it are significant for woman not being selected to job interviews because of the biases sustained by AI. The structural injustice is, therefore, combined with discrimination because they are both the effect of the existing prejudices in the society. Secondly, AI strengthens also the epistemic injustice which is based on the assumption that in the society already exist the patterns of identifying people with concrete jobs, behaviours or skills, which are the consequence of racial and gender biases such as 'looking like man' which is identified by AI with 'looking like a White man' (Rafanelli 2022: 4; Wittkower 2018). Rafanelli emphasises that the abovementioned injustices disclose the problem of power hidden behind the AI technology. 'My point is that AI is a tool with which humans exercise power, rather than a replacement for human power, and its use therefore raises questions of justice. It is our responsibility as consumers, programmers, and researchers to ensure these questions don't go unanswered' (Rafanelli 2022: 5).

In order to apply the above argument to the analysis of AI, a solution must be found to change the algorithmic discourse by controlling its normative content. There are solutions which might be introduced to AI to guarantee that the normative effects of algorithmic performative power of AI would be more just and normatively accepted. The first one is connected with the guaranteeing of diversity among designers and programists (Waelen and Wieczorek 2022) which can ensure greater sensitivity of algorithms to gender, racial, ethnic issues, etc. The second one is based on the change in the data which are used to train AI—if there are no biased real data, there might be used artificial data but prepared in a way which will be neutral from a normative point of view (Jacobsen 2023). The third solution is to conduct algorithm audits to monitor their operation, without disclosing the corporation's intellectual property, and fourth, educate people more broadly on how algorithms work (Burrell 2016). However, all these solutions suggest that AI should be treated as an epistemic tool rather than normative—embedding or delegating into this technology a value system that prevents it from making errors in the identification and categorisation of people is not, according to me, equal to delegating a normative power to AI similarly to e.g. speed bumps (Latour 1999) because it does not affect *human behaviour* but only the *technical operation of AI*. Moreover, the fact that there might be no misrecognition in the work of AI in a normative sense does not necessarily mean that there occurs recognition in the normative sense. The problem of the normative consequences of AI is much more visible when it makes mistakes, but is flawless recognition of the quality of things by AI immediately equivalent to recognition in the normative sense? And does every error in identification and categorisation made by AI necessarily lead to a 'struggle for recognition' in the normative sense? Is it not a question of a specific form of epistemic representation of individuals and social groups, of how they understand and know themselves, also in confrontation with AI algorithms?

If AI repeats prejudices, we have to struggle for change in the perspective of that how different groups of people are represented in the social imaginary, not only through the algorithms. In this sense, any errors made by AI in identifying people, which can result in the misrecognition of individuals and thus have normative consequences, also provide an opportunity to question what form of recognition we expect from this technology. Assuming that AI will flawlessly identify human faces and also accurately diagnose people's emotions, health status, likelihood of committing an offense, or ability to repay a loan, will lead to the vision of the world with no escape from surveillance capitalism (Zuboff 2019) or a datafied society (Kitchin 2022). 'And AI-based facial recognition technology is often used for surveillance and can violate people's privacy. (…) No information from your phone and no biometric data are needed. The machine does its work from a distance' (Coeckelbergh 2020, 6). Artificial intelligence could not only be a tool to assist in complex computational analyses but also a basis for making specific, possibly also socially, ethically and politically-sensitive, decisions (Jefferson 2018; Eubanks 2018;

Dencik et al. 2019). On the one hand, it is clear that AI's recognition or misrecognition of an individual or group can affect the respect for their dignity and the conditions for the formation of their identity in the similar meaning of recognition to Taylor's theory, which requires a solution at the level of the technical capabilities of these devices. On the other hand, there is a growing concern that perfect identification of individuals by AI tools will lead people to being locked into a cage of diagnosed identities and assigned categories. The fact that AI can influence users' self-respect and integrity, perpetuate stereotypes, polarize society, or reproduce epistemic injustice suggests that the proper identification of users, free from biases, is a normative requirement for AI. However, algorithms refer to our data, not to ourselves (Cheney-Lippold 2017), and reminding this fact may broaden our perspective on (mis)recognition by AI. In other words, recognition or misrecognition, in the context of AI, may refer to a specific representation of ourselves constructed by AI on our data. The question that must therefore be asked is: how does its algorithmic representation, our data doubles, affect our identity? I answer this question in the next section.

4 Datafied Concept of Identity

Cheney-Lippold proposes a closer look at what AI 'talks' about us and why opening ourselves up to situations in which misrecognition may sometimes occur does not necessarily have such clearly negative consequences for the subject. He reminds us that algorithms diagnose not people and their personality, but data. Algorithms diagnose data, not specific individuals, which means that identification and categorisation by AI simply means that certain data are classified into specific categories. To put it simply, the fact that a user of a given application, browser, or social media platform stops scrolling posts in a particular moment, searches for specific information, or watches a particular video is considered part of their identity, as important information about who they are, even though, in reality, all this data builds, at most, their online persona and does not necessarily tell the truth about them as a specific person. 'Websites will take the fact that you went to a certain site as evidence of your identity' (Cheney-Lippold 2017, 7), which essentially boils down to the concept of identity as something built on data rather than qualitative experiences, emotions, or reflections.

The way in which AI identifies and categorises data is based on identifying statistical patterns on a large amount of data from different users that allows a specific person to be matched to a so-called 'measurable type'. 'A measurable type is a data template, a nexus of different datafied elements that construct a new, transcoded interpretation of the world' (Cheney-Lippold 2017, 47). What needs to be distinguished in interactions with AI in their conceptual, cognitive, and ethical dimensions is the logical level of assigning specific data to specific categories. In other words, as a particular person, I can identify myself as a woman, a philosopher, a rock music listener, and a cat lover, but for AI, my identity breaks down into categories: 'woman', 'philosopher', 'rock fan', 'cat lover'. These quotation marks play a key role, representing categories to which many other specific pieces of data about individuals or groups also belong, according to an AI algorithm. Although the categories used to classify me seem broad enough to encompass my self-identification, it would be inaccurate to suggest that these categories alone define my identity.

> Our datafied lives, when aggregated and transcoded into quotation-marked categories, increasingly define who we are and who we can be. But to make sense of this aggregation requires a new definition of 'man,' one based on website visits instead of performance or self-identification (Cheney-Lippold 2017, 19).

If we reduce a person's identity to what they do online, what content they view, and what apps they use, then we are simply using a new, quantified and datafied concept of identity. Of course, considering how much time we spend online nowadays, it may turn out that our understanding of ourselves is indeed significantly shaped by the representation of ourselves available to us through AI tools. Cheney-Lippold argues, however, that in this case, we must perceive situations of recognition or misrecognition not so much as the consequence of a well- or poorly-functioning technology but as a new form of categorisation of the subject. Recognition or misrecognition in the algorithmic language means that 'algorithmic "race" and "gender" isn't about being a white man. It's about being a "Caucasian", "man" with a confidence measure of 87 percent' (Cheney-Lippold 2017, 34). The fact that our data is defined as a percentage match for specific categories indicates that the AI's interpretation of us is statistical in nature and based on recognition in the epistemic sense. The AI tool categorises our data in this way, and whether we consider this categorisation to be correct or not depends on the individual or group, or more broadly, mechanisms of power, when AI-generated content is considered to adequately represent social reality.

> Correspondingly, in an online world of endlessly overhauled algorithmic knowledge, Google's misrecognition of my gender and age isn't an error. It's a reconfiguration, a freshly minted algorithmic truth that cares little about being

> authentic but cares a lot about being an effective metric for classification. In this world, there is no fidelity to notions of our individual history and self-assessment (Cheney-Lippold 2017, 9).

The perspective adopted by Cheney-Lippold allows us to see the difference between recognition in the epistemic sense and in the normative sense. Technology such as AI does indeed recognise specific data and classify it, but this operation is strictly epistemic in nature—it is making something recognisable (Butler 2021b). These classifications only acquire normative value when they are interpreted as recognition or misrecognition. In other words, as a user of a particular application, I may not even know how the algorithm categorises me. However, when I realise that as a female rock fan I have been classified as a listener of 'dad rock', I can interpret this classification as misrecognition, even if an algorithm might correctly interpret my musical taste. The fundamental problem with AI algorithms is that, as users of this technology, we have no influence on how it categorises our data. The fact that AI constructs categories, which we then consider to be significant for our identity (and, more broadly, having social and political consequences), is a normative consequence of AI's epistemic recognition. However, these categories are not normative by definition. The categories 'human being', 'race', and 'rock fan' have no ethical value until they are placed in quotation marks. However, when they become an accurate or inaccurate description of reality and justify specific actions based on these categorisations, they become morally relevant.

> (…) a sense that although we alone don't make up 'gender,' 'gender' affects us all. You, as an individual, might not want to be profiled and manipulated without your knowledge, but your fate is tied to the rest of the population whose dividual lives become the patterns by which you are recognized (Cheney-Lippold 2017, 238).

Misrecognition, in that case, cannot be avoided, as AI does not refer to individuals in a personal sense, but to patterns derived from data. In this sense, AI is another medium with hermeneutic significance—the meaning of categories that we discover through algorithms may differ significantly from our sense of self, both individually and collectively, but it may not only constitute misrecognition in a negative sense but also serve as a basis for interpreting who we are in our data double (see Juchniewicz and Wieczorek 2024). 'We are thus made subject not to our data but to interpretations of that data' (Cheney-Lippold 2017, 258).

If, following Butler, we assume that the conditions of recognizability precede recognition in the normative sense, as they constitute the very framework in which recognition takes place, then AI algorithms are precisely such a digital framework of recognition. We are left to ask ourselves whether the world represented by AI categories is entirely about us, or rather about a certain metanarrative about ourselves represented in the data that is subjected to AI categorisation. The fact that AI can be a tool for self-interpretation reminds us that the essence of recognition is not only about building relationships with others, but also about turning inward. In the next section, I analyse what this inward turn might mean in the context of human-AI relationships.

5 AI, Reflexivity and Partial Recognition

As a user of AI-assisted apps, I can do something completely different with the classification of my data conducted by this technology in comparison to treat them as just expressing (or not) recognition of my person, namely I can reflect on what it actually means for my identity. I can 'return to myself' (Hegel 1977), which is ultimately what recognition is all about. If I personally feel or see that the algorithm has incorrectly categorised me into a certain group of users, I can treat this error as a significant element of my identity. If the type of music I like to listen to is mostly listened to by middle-aged fathers, I can, for example, notice that my identity has also been strongly influenced by male music culture; that I belong to a specific generation that associates certain memories with the important role of a particular musical period, or that I simply listen to music that connects me emotionally with my own father (if I am a child and listen to 'dad music'). Our identity is much more ambiguous than we think. If the AI algorithm determines that my skin colour is only 87% white, if only 45% of my interests match those of a woman because my hobbies are fishing and playing golf, which are dominated by men, if I work in a field highly represented by older people, which is why Google thinks I am 65 (even though I am 40), it shows that my identity is much more fluid, queer, and trans-age. The above examples do not, of course, refer to serious violations of civil rights and freedoms that may occur in the case of errors in identifying people, as it was explained in Sect. 3. At the level of the self-identity of individuals or social groups, misrecognition does not, however, necessarily imply disrespect for specific individuals in a normative sense. Moreover, if we assume that recognition is a relationship built not only with the Other, but also a reflexive relationship aimed at self-knowledge, then confrontation with non-obvious categorisations of ourselves can have a significant impact on our understanding of the complexity of our own identities.

Butler suggests that recognition is not a condition which can be achieved once by a person in the society but rather is 'an ongoing dynamic in social relations, one that does not come to a teleological fulfilment at any certain time' (Butler 2021b, 27). The dynamic dimension of recognition indicates that recognition should not be understood as 'full', 'closed' or 'achieved', because it is almost never possible to be fully recognised on all levels of personal needs, aspirations and self-development. Butler suggests, therefore, that recognition is always partial:

> (…) there is, for instance, no recognition of all that a person is or may be, and in that sense, there is a capture of the person by existing terms available for the conferring of recognition. But the fact that recognition is partial does not make it false. It only makes it noncomprehensive (Butler 2021b, 27).

Following Butler's assumption that in social life no one ever achieves full recognition, because we ourselves do not have insight into our full identity, how we understand ourselves and how we want to be understood by other people, it becomes clear that expecting full recognition from AI is equally impossible. AI algorithms refer to our data, not to us, and although I may want to train an algorithm to best reflect my real interests or expectations regarding recommended content and information, it is always my interpretation of what AI recommends to me. It is I who recognise, in an epistemic sense, that the algorithm of a particular application works in accordance with my sense of self and what interests me, what values I agree with, and what content I want to interact with.

The approach that recognition could be partial is important for problems raised in the research of AI since it delivers a possibility to see that AI may not necessarily fully recognise people by a proper identification of them, however, could play a significant, partial or distributive role in this process of identification having significant normative consequences (Juchniewicz 2024). Even if practices of love, law or solidarity, important from Honneth's perspective, do not undergo a radical change in meaning in society due to the existence of AI, for individual users, the fact that they can engage in practices of care, perform work or ask questions about freedom and justice in confrontation with technology is a change that allows them to reflect on both specific practices and their meaning anew.

Airoldi states that:

> Rather than perfectly meeting or nudging their users' needs and wants (...) machine learning systems negotiate with them a common cultural ground, iteration after iteration, feedback after feedback. The sum of these non-linear trajectories ultimately brings about the techno-social reproduction of society (Airoldi 2022, 132).

The reproduction of social order in the logic of algorithms is not a linear process, as modern algorithms work non-linearly. They learn to recognise data, and when they make mistakes, they try to correct them and learn what the mistake was. AI may therefore constitute a new medium for transforming the meanings of specific concepts and categories. Until now, as Butler emphasises, this role has been played by social movements, which have overtaken the specific meanings of the categories in which individuals and groups were defined and given them new meanings (Butler 2021b, 37). To a certain extent, this is now also possible thanks to AI, which can not only limit the impact of social movements but also strengthen and support them.

6 Conclusion

In this chapter, I asked whether artificial intelligence can be a carrier of recognition and whether it is possible to shape individual and collective identity through it. In Sect. 2, I explained why the concept of recognition is important from the normative perspective and showed, using the theories of Honneth and Taylor, that both individual and collective identities depend on recognition or misrecognition. I compared this normative orientation in recognition theory with the perspective emphasising the epistemic dimension of recognition represented by Butler. Butler raises the issue of the very conditions that define recognizability and links them to the interpellative and performative power of language and discourse. Both perspectives on the meaning of recognition, normative and epistemic, intersect in AI research.

In Sect. 3, I presented how normative problems related to the use of AI are diagnosed and why they affect the recognition that users of this technology may experience. I also showed the epistemic power of AI, which makes this technology an important carrier of recognition. However, I pointed out that expecting AI to be fully technically effective in granting recognition may pose a threat from a political and social perspective. At the same time, I noted that misrecognition is not always an actual problem in the normative sense and, moreover, that there are many situations where recognition in the technical sense occurs without users being aware of it, so this form of recognition does not necessarily have specific moral consequences. That is why I proposed to look at AI from the point of view of epistemic recognition, and not necessarily only normative one.

Following this line of thought, in Sect. 4, I showed that identity built in relation to algorithms requires users of this technology to realise that AI diagnoses not them but their data. Furthermore, the categories used by AI are not normative in nature—in other words, it is difficult to accuse technology of 'wanting' to treat us in a particular way or harm us by assigning us certain characteristics or qualities that it recognises in large data sets. However, this does not change the fact, as Cheney-Lippold (2017) argues, that if we assume that the world represented in AI categories is an adequate way of describing the social world, then the way in which AI presents content to us will undoubtedly have a significant impact on the conditions in which our identity is formed. In other words, AI first interprets data and then reproduces these interpretations when users do not attempt to change them in any way. Recognition within 'human-AI' relationships should be understood as 'my data-AI', and thus an important identity issue is the extent to which our data doubles can serve as a reference point for how we perceive and understand ourselves. AI is an important hermeneutic tool through which we can interpret ourselves.

In Sect. 5, I therefore drew attention to the reflexivity that should be present in a relationship of recognition. One of the important conclusions of Butler's theory is that it is impossible to expect full recognition in social life—it is always partial or distributed. Thus, the expectation that technology will recognise us perfectly may also be a blind spot in considerations about AI. Our interactions and relationships with AI show rather how broad the framework for shaping our identity is, since it can be influenced by algorithms, the categories in which AI sees us, or simply the technologies we use to carry out our daily activities.

When applying theories of recognition to AI and its influence on identity formation, we must not only emphasise the normative dimension of recognition, but also the epistemic framework that determines what is recognisable. As Butler argues, our identity is shaped not only by specific practices that carry certain moral consequences, but also by meanings, concepts and discourses that define the framework of our normative recognition practices. Through the categorisation of user data, artificial intelligence today determines precisely such an epistemic framework for normatively sensitive practices, thereby helping to co-shape the conditions for the development of our identity. Just like the normative recognition that we experience in our interaction with AI technologies can impact identity formation, the epistemic recognition represented in AI's categorisation and the data doubles it creates of us can also shape a person's identity. The epistemic recognition of AI can nudge us to reflect on our own identity and features and as such shape and change our self-understanding. AI is proving to be an important hermeneutic tool for our individual and collective identity.

References

Althusser, Louis. 2001. Ideology and Ideological State Apparatuses. In *Louis Althusser, Lenin and Philosophy and Other Essays (Transl. B. Brewster)*, 122–174. New York: Monthly Review Press.

Airoldi, Massimo. 2022. Machine habitus: Toward a sociology of algorithms. Cambridge: Polity.

Badarin, Emile. 2023. Politics of Recognition, Elimination and Settler-Colonialism. *Critical Sociology* 49 (2): 233–252. https://doi.org/10.1177/08969205211058452.

Balaton-Chrimes, Samantha, and Victoria Stead. 2017. Recognition, Power and Coloniality. *Postcolonial Studies* 20 (1): 1–17. https://doi.org/10.1080/13688790.2017.1355875.

Burrell, Jenna. 2016. How the Machine 'Thinks': Understanding Opacity in Machine Learning Algorithms. *Big Data and Society* 3 (1): 2053951715622512. https://doi.org/10.1177/2053951715622512.

Butler, Judith. 2021a. *Excitable Speech. A Politics of the Performative*. London, New York: Routledge.

Butler, Judith. 2021b. Recognition and the Social Bond: A Response to Axel Honneth. In *Recognition and Ambivalence: Judith Butler, Axel Honneth, and Beyond*, edited by Heikki Ikäheimo, Kristina Lepold, and Titus Stahl, 31–54. New York: Columbia University Press.

Chari, Anita. 2004. Exceeding Recognition. *Sartre Studies International* 10 (2): 110–122.

Cheney-Lippold, John. 2017. *We Are Data. Algorithms and the Making of Our Digital Selves*. New York: New York University Press.

Coeckelbergh, Mark. 2020. *AI Ethics. Cambridge, Massachusetts, London, England: The MIT Press.*

Dencik, Lina, Arne Hintz, Joanna Redden, and Emiliano Treré 2019. Exploring Data Justice: Conceptions, Applications and Directions. Information, *Communication & Society 22*(7): 873–881.https://doi.org/10.1080/1369118X.2019.1606268

Eubanks, Virginia. 2018. *Automating inequality: How High-Tech Tools Profile, Police, and Punish the Poor*. St. Martin's Press.

Fraser, Nancy, and Axel Honneth. 2003. *Redistribution or Recognition? A Political-Philosophical Exchange*. London: Verso.

Fricker, Miranda. 2018. Epistemic Injustice and Recognition Theory: A New Conversation —Afterword. *Feminist Philosophy Quarterly* 4 (4): https://doi.org/10.5206/fpq/2018.4.6235.

Hegel, G. W. F. 1977. *Phenomenology of Spirit (Trans. A. V. Miller)*. Oxford, New York, Toronto, Melbourne: Oxford University Press.

Hegel, G. W. F. 1983. *Hegel and the Human Spirit. A Translation of the Jena Lectures of Spirit (1805–6) with Commentary (Trans. Leo Rauch)*. Detroit: Wayne State University.

Hirvonen, Onni. 2017. Groups as persons? A suggestion for a Hegelian turn. *Journal of Social Ontology* 3 (2): 143–165.

Honneth, Axel. 1995. *The Struggle for Recognition*. The MIT Press. https://mitpress.mit.edu/9780262581479/the-struggle-for-recognition/.

Honneth, Axel. 2021. Recognition Between Power and Normativity. A Hegelian Critique of Judith Butler. In *Recognition and Ambivalence: Judith Butler, Axel Honneth, and Beyond*, edited by Heikki Ikäheimo, Kristina Lepold, and Titus Stahl, 21–29. New York: Columbia University Press.

Ikäheimo, Heikki. 2002. On the Genus and Species of Recognition. *Inquiry* 45 (4): 447–462. https://doi.org/10.1080/002017402320947540.

Ikäheimo, Heikki, and Arto Laitinen, eds. 2011. *Recognition and Social Ontology*. Leiden: Brill.

Ikäheimo, Heikki, Kristina Lepold, and Titus Stahl. 2021. Introduction. In *Recognition and Ambivalence: Judith Butler, Axel Honneth, and Beyond*, edited by Heikki Ikäheimo, Kristina Lepold, and Titus Stahl, 1–20. New York: Columbia University Press.

Jacobsen, Benjamin N. 2023. Machine Learning and the Politics of Synthetic Data. *Big Data and Society* 10 (1): https://doi.org/10.1177/20539517221145372.

Jefferson, Brian Jordan. 2018. Predictable Policing: Predictive Crime Mapping and Geographies of Policing and Race. *Annals of the American Association of Geographers* 108 (1): 1–16. https://doi.org/10.1080/24694452.2017.1293500.

Juchniewicz, Natalia. 2024. Wie künstliche Intelligenz die Theorie Der Anerkennung verändert. In *Philosophische Digitalisierungsforschung: Verantwortung, Verständigung, Vernunft, Macht*, edited by R. Adolphi, S. Alpsancar, S. Hahn, and M. Kettner, 115–136. Transcript.

Juchniewicz, Natalia, and Michał Wieczorek. 2024. Self-Tracking, Background(s) and Hermeneutics. A Qualitative Approach to Quantification and Datafication of Activity. *Phenomenology and the Cognitive Sciences* 23:133–154. https://doi.org/10.1007/s11097-022-09821-x.

Kislev, Shachar F. 2020. Six Hegelian Theses About Technology. *Techné: Research in Philosophy and Technology* 24 (3): 376–404. https://doi.org/10.5840/techne2020730125.

Kitchin, Rob. 2022. *The Data Revolution. A Critical Analysis of Big Data, Open Data and Data Infrastructures*. Los Angeles, London, New Delhi, Singapore, Washington DC, Melbourne: Sage.

Laitinen, Arto. 2002. Interpersonal Recognition: A Response to Value or a Precondition of Personhood? *Inquiry* 45 (4): 463–478. https://doi.org/10.1080/002017402320947559.

Latour, Bruno. 1999. *Pandora's Hope: Essays on the Reality of Science Studies*. Cambridge, Massachusetts: Harvard University Press.

Mcintosh, Monique. 2022. Does the Politics of Recognition Function as a Mechanism for the Amelioration of Colonialism's Effects, or as a Means through Which These Effects Are Reproduced? *Kent Law Review* 7 (1): https://doi.org/10.22024/UniKent/03/klr.1091.

Rafanelli, Lucia M. 2022. Justice, Injustice, and Artificial Intelligence: Lessons from Political Theory and Philosophy. *Big Data and Society* 9 (1): 205395172210806. https://doi.org/10.1177/20539517221080676.

Taylor, Charles. 1994. The Politics of Recognition. In *Multiculturalism: Examining the Politics of Recognition*, edited by Amy Gutmann, 25–73. Princeton: Princeton University Press.

Waelen, Rosalie A. 2022. The Struggle for Recognition in the Age of Facial Recognition Technology. *AI and Ethics* 3 (1): 215–222. https://doi.org/10.1007/s43681-022-00146-8.

Waelen, Rosalie, and Michał Wieczorek. 2022. The Struggle for AI's Recognition: Understanding the Normative Implications of Gender Bias in AI with Honneth's Theory of Recognition. *Philosophy and Technology* 35 (2): 53. https://doi.org/10.1007/s13347-022-00548-w.

Wittkower, D. E. 2018. Discrimination. In *Spaces for the Future. A Companion to Philosophy of Technology*, edited by Joseph C. Pitt and Ashley Shew, 14–28. New York, London: Routledge.

Zuboff, Shoshana. 2019. *The Age of Surveillance Capitalism: The Fight for a Human Future at the New Frontier of Powe*r. New York: Public Affairs.

Design for Recognition and Its Dangers

Nynke van Uffelen and Fernando Secomandi

Abstract In this chapter, we explore opportunities and pitfalls when adopting recognition theory as a normative paradigm for technology design. Building on Honneth's notion of justice as recognition, we characterise technologies as constituted by social relations of (mis)recognition, and people's identities as relationally constituted in processes mediated by technologies. As such, we argue, people can (mis)recognise each other *through* technologies, shaping—or failing to shape—undistorted practical self-relations and autonomy. This perspective on 'recognition through technology' emphasises that technologies are always *designed*, implying that identities too are, at least in part, designed. As such, we introduce the notion of *design for recognition*, and we explore its moral appeal, merits, and dangers. We argue that although design for recognition offers conceptual, normative and empirical advances to postphenomenological mediation theory, one of the dominant approaches in

N. van Uffelen (✉)
Technology, Policy and Management, Delft University of Technology, Delft, The Netherlands
e-mail: n.vanuffelen@tudelft.nl

F. Secomandi
Technology, Policy and Management, Delft University of Technology, Delft, The Netherlands

Faculty of Industrial Design Engineering, Delft University of Technology, Delft, The Netherlands
e-mail: f.secomandi@tudelft.nl

R. Waelen (ed.), *Normative Relations of Recognition in the Age of AI*, Philosophy and Ethics of AI 2, https://doi.org/10.1007/978-3-032-34028-3_5

philosophy of technology, there are four dangers to designing for recognition that should be considered, namely: (a) moral intuitions can be distorted; (b) structural injustices may be (re)produced or obscured; (c) value hierarchies and social norms may become dominating once embedded in technologies; (d) identities may be reified fuelling political polarisation. The introduced vocabulary around 'design for recognition' in this chapter sets an interdisciplinary and empirically-informed research agenda for critical theorists, philosophers of technology, and designers.

Keywords Design ethics • Post-phenomenology • Human-technology relations • Recognition theory • Misrecognition • Philosophy of technology

1 Introduction

Critical theorists aim to formulate well-grounded societal critiques to achieve emancipation and social justice, and in doing so, they value being interdisciplinary and empirically informed (Thompson 2017). Much recent work in critical theory revolves around the notion of recognition, inspired by Axel Honneth, who conceives injustice in terms of relations of misrecognition (Honneth 1995). However, despite the fact that contemporary social relations are increasingly entangled with concrete technologies, the possibility of technology-mediated recognition has largely escaped the critical theorists' gaze. More specifically, emerging technologies, including artificial intelligence (AI), and their applications in fields such as medicine, government, and energy infrastructure, increasingly mediate relations of (mis)recognition, shaping people's identities in fundamental ways. Recognition theory, in particular, contains conceptual resources to analyse and normatively evaluate these processes (Gertz 2018; Joel Anderson 2025; Secomandi and van Amstel 2025; van Uffelen 2022; Waelen 2022; Waelen and Wieczorek 2022).

In this chapter, we seek to contribute to this line of research by exploring opportunities and pitfalls when adopting 'justice as recognition' as a normative paradigm for the constitution of people's identities through technology design. We start in Sect. 2 by arguing for an interpretation of 'recognition through technology' as implying that social relations of (mis)recognition and identity formation based on love, right, and esteem run not only through traditional institutions such as the economy, laws, and education, but also through concrete technologies. Additionally, drawing from the philosophy of technology—in particular from the postphenomenological line of research developed by Don Ihde (1990), Peter-Paul Verbeek (2011), among others—we argue that human experiences of recognition, and therefore the formation of their identities, can also be technologically mediated in significant ways. Additionally, because technologies do not come into being if not through a *design* process, we conclude that social identities are, at least in part, designed.

In Sect. 3, we explore further the notion of 'design for recognition' as a postphenomenological basis for technology design. We dive deeper into Verbeek's framework of 'technology-mediated morality' and incorporate Secomandi and van Amstel's (2025) revision of that framework, inspired by Hegel's recognition theory, to flesh out how design can influence identity formation in an intersubjective fashion.

Then, in Sect. 4, we discuss some risks of designing for recognition, inspired by recent authors outlining ambivalent aspects of the recognition concept (Laitinen 2021; Stahl et al. 2021). We argue that there are four main dangers to designing for recognition that should be considered, namely: (a) moral intuitions can be distorted; (b) structural injustices may be (re)produced or obscured; (c) value hierarchies and social norms may become dominating once embedded in technologies; (d) identities may be reified, fuelling political polarisation.

Introducing a conceptual repertoire around 'design for recognition' emerging from the notion of 'recognition through technology' gestures toward an ambitious research agenda in a range of disciplines. Philosophers of technology may explore recognition theory as a useful lens, as its added value, we will argue, is conceptual, empirical, and normative. Next, the vocabulary can enrich the critical-reflexive practices of designers. Lastly, we hope that critical theorists start exploring the social and political systems underlying concrete empirical technologies, enriching their analysis and expanding the unit of social critique.

2 Recognition Through Technology: The Centrality of Design

To explain our interpretation of 'recognition through technology', we first elaborate on the meaning of 'recognition', drawing on Axel Honneth's works, especially his *The Struggle for Recognition* (Honneth 1995). Ultimately, the question Honneth is committed to answering with his book, in line with the Frankfurt School, is the following: how to ground normative societal critique? To do so, his phenomenological point of departure is collectively shared experiences of disrespect or injustice, which are immediately felt by people as a violation of social norms or expectations (Honneth 1982). As such, experiences of injustice bolster a normative potential for emancipation.

Honneth understands social struggles for emancipation and justice as struggles for recognition, rooted in experiences of misrecognition. Adapting

Hegel's three spheres of social life—family, civil society, and the state—he distinguishes between recognition through love, solidarity, and rights. Love provides the foundation of self-confidence, while its distortions yield damaged self-relations. Solidarity rests on shared cultural values that attribute esteem to different forms of activity, shaping individuals' sense of self-worth. Rights secure dignity and self-respect by affirming individuals as autonomous and equal, with their denial producing experiences of disrespect.

Three important aspects within this theory are relevant to our aims in this chapter. First, relations of recognition do not occur between already fully-formed people. On the contrary, they *constitute* people's very autonomy and sense of personal identity. Genuine relations of recognition lead to practical relations-to-self (e.g., self-confidence, self-esteem, and self-respect) that are undistorted, while misrecognition generates harmful self-relations, harming an individual's autonomy (Anderson and Honneth 2009). In other words, the self is intersubjectively constituted in a fundamental way; however, this process may be corrupted. Honneth takes a strong normative stance here, defending that all people need and deserve undistorted self-relations, and therefore, misrecognition equals injustice. Second, all three relations of (mis)recognition not only exist at the interpersonal level but are to a large extent *institutionalised*. For example, through the legal system, people may or may not be recognised as citizens with rights (e.g., human rights, anti-discrimination laws). Through the market, people may or may not be recognised as contributors, whether through employment, wages, or professional status. These institutions encode, distribute, and regulate recognition beyond the interpersonal level. Third, and as a consequence of the above, recognition provides a normative standard against which institutions can be judged. Institutions that obstruct the full development of individuals' autonomy and self-relation can therefore be denounced as unjust.

Given this understanding of recognition, the link with technology becomes fairly straightforward. If technologies, including concrete objects, can be seen as artefacts created and used within an institutional context in which people relate to one another, then they can also configure relations of (mis)recognition. In other words, technologies can mediate relations of recognition between human beings. Although Honneth has not yet, to our knowledge, elaborated on this, his perspective—including his three forms of recognition—seems compatible with the possibility of technologically mediated recognition. Recently, philosophers of technology have started to explore relations of misrecognition influenced by technologies that result in problematic formations of personal identities (Gertz 2018; Secomandi and van Amstel 2025; van Uffelen 2022; Waelen 2022; Waelen and Wieczorek 2022).

Ultimately, however, technologies are never naturally given; they are created by people (van de Poel 2021; van de Poel and Kroes 2014; Verbeek 2006; Winner 2017).[1] Thus, people's identities, including their practical relations-to-self, mediated by technologies, are at least partly constituted, or *designed*, by others. Next, we turn to how this claim can be corroborated by ongoing developments within the philosophy of technology.

3 Identities Are Designed: A Dialectical-Postphenomenological Perspective

Given that misrecognition is unjust, and that people can misrecognise each other through technologies, we can conclude that there is a real moral imperative to avoid inscribing relations of misrecognition in technologies through design. The responsibility to do so falls primarily on designers and those socially empowered to shape concrete artefacts and systems to be used by others. Therefore, we propose the notion of 'design for recognition', which denotes that recognition should be taken into account when developing technologies.

Analyses of misrecognition through technologies can focus on at least three sources: (a) the design process; (b) the 'material' form of technologies; and (c) the effects technologies have on users. In the design process, people can be misrecognised as their voices or input are excluded, marginalised, dismissed, or altogether ignored due to being, for example, of a certain race, gender, or religion (Fraser 2000; Fricker 2007, 2018). Next, relations of misrecognition can be embedded in the objective configuration of technologies, including physical materials and shapes, but also model algorithms and code. For example, a language model trained on biased datasets can be experienced as producing sexist outputs in interaction with users. Third, technologies have effects in the context in which they are used, affecting power relations, generate income, profile prospective criminals, or fuel war. Some of these effects are distributive, some are cultural, yet all these effects may affect people's practical relations-to-self, thus leading to misrecognition of social groups (Fraser and Honneth 2003).

[1] In Chaps. 9, 10 and 11 of this book, authors debate whether technologies can (mis)recognize people autonomously, without a human 'in the loop'. In this chapter, we sideline this possibility and focus on people-people (mis)recognition through technologies.

For example, relations of *love* may be mediated through technologies such as care robots for the elderly, influencing these users' perceptions of what it means to be lonely and dependent on others in contemporary societies. Similarly, relations of *solidarity* may be distorted by algorithmic ranking systems on social media that privilege certain forms of self-expression over others, thereby shaping who is deemed worthy of esteem in a given community. Likewise, relations of *rights* can be undermined when automated welfare systems wrongly flag citizens as fraudulent claimants, effectively denying them recognition as equal members of a political community. In each of these cases, the technologies in question mediate identity formation by shaping whether individuals can develop undistorted self-relations.

The notion that social relations embed and help to define technologies and are, in turn, shaped by them is not new within the philosophy of technology. For example, the concept of *sociotechnical systems* in science and technology studies (STS) stresses that technologies contain both technical and social aspects and are 'socially constructed and society shaping' (Thomas P. Hughes 1987, p. 1). Alternatively, *design for values* or *value-sensitive design* approaches focus on how values (should) shape and constitute the design of technologies (van de Poel 2021; van de Poel and Kroes 2014). Moreover, mediation theory, another prevalent perspective in the field, emphasises the role of technologies in mediating human-world relations (Ihde 1990), thereby shaping moral subjectivities (Verbeek 2011). All such perspectives share a relational and constructionist ontology of technology, acknowledging it as both constituted by and constitutive of people's practices in a socially shared world.

To what extent, then, is design for recognition a valuable addition to this discussion? In answering this question, we focus on postphenomenology, one of the leading strands of contemporary philosophy of technology (see Achterhuis 2001; Mitcham 1994; Olsen et al. 2009). We draw primarily on Secomandi and van Amstel's (2025) Hegel-inspired dialectical revision of the concept of technological mediation. In their article, the authors rework Verbeek's (2011) framework of *mediated morality*, which marks postphenomenology's seminal incursion into the theme of design and its role in shaping moral subjectivity. In what follows, we identify three key contributions from their perspective that offer conceptual, normative, and empirical grounding for what we call 'design for recognition'.

First, the concept of recognition through technology expands the *conceptual* scope of mediation theory. The authors observe how Verbeek's original framework posits a relational ontology in which morality is always co-constituted by humans and nonhumans. Specifically, his concept of 'relational freedom' shifts the focus from autonomy as an achievement of individual humans to a distributed process emerging through their interactions with technological artefacts. Drawing on Foucault's notion of subjectivation, Verbeek further proposes that humans are constituted as moral subjects through their practical engagements with concrete technologies. Freedom, then, is not freedom from mediation, but freedom *in* and *through* mediation. This framing helps illuminate how technologies not only co-shape how the world is experienced by humans but, in the process, also influence moral subjectivity itself.

While this is a crucial insight, Secomandi and van Amstel argue that Verbeek's framework tends to treat humans as isolated individuals forging their moral subjectivity in interaction with technologies, without adequately experiencing the social underpinning of this process. In contrast, a recognition-based approach builds on the Hegelian notion that human subjectivity is inherently dependent on experiencing *others*: to become a self-conscious being, one must both recognise and be properly recognised by another self-consciousness. From this perspective—and in alignment with our aim of expanding Honneth's earlier analyses—, human subjectivity can be mediated not only by technological artefacts but also through relations with other humans, who are themselves technologically mediated. Thus, recognition theory can complement and extend postphenomenology by providing a dialectical account of technologically mediated relations between self and other. It helps to show that human-technology relations that is not only *interobjective* (i.e., between human subjects and nonhuman objects) but also *intersubjective* (i.e., between subjects), attuned to how identities are shaped in socially embedded, technologically mediated contexts.

Second, recognition theory reinforces the *normative* ambition of technological mediation theory. Verbeek acknowledges that technologies 'materialise' morality and that people bear responsibility for shaping technological mediations through the forms they give to concrete artefacts. However, it is unclear whether mediation theory contains sufficient resources to elucidate when technological mediation can be considered desirable or not.[2] Verbeek rightly recognises the inalienable agency of humans and their capacity to 'style' their own moral subjectivity through nonhuman technologies. Also, he defines freedom as a relational concept and as an outcome of technological mediations, which are always shaped by human involvement. However, he limits this process to the interaction between individuals and technologies, omitting the connection that these people can establish with the intentions of human

[2] For an attempt to develop a critical theory of postphenomenology, see Feenberg (2020); for a similar argument in the context of value-sensitive design, see Jacobs and Huldtgren (2021).

others and, more specifically, the ones who originally gave form to the technologies. As such, his framework is unable to reveal how technology can feel oppressive for some due to the way it was materialised by others, whether with deliberate intention or not.

Hegelian recognition theory offers a remedy by conceptualising freedom as a relation between self-determining, reciprocally recognised beings. In terms of the Honnethian perspective adopted in this chapter, this translates into undistorted self-relations, which arise only when individuals are adequately recognised by others in the domains of love, solidarity, and right. Misrecognition through technology, then, is unjust insofar as it distorts an individual's practical self-relations within an established context of mutual dependence on other human beings. This insight allows us to articulate a normative standard for evaluating technologies: when used, do they facilitate or obstruct the recognition of people's identities in ethical relations to other social agents?

Third, recognition theory can aid *empirical* analysis within design processes in two ways. On the one hand, it provides researchers and designers with concepts, typologies, and distinctions to analyse experiences of misrecognition and distorted practical self-relations (for such a typology, see van Uffelen and Santos Ayllón 2025). A thorough problem analysis, one that outlines the types of misrecognition experienced in concrete empirical contexts, can contribute to more nuanced technology critiques and strategies for mitigation and emancipation through design. On the other hand, recognition theory offers a valuable lens through which to empirically analyse the ethical role of designers in shaping technological mediations. Verbeek asserts that designers inevitably influence users' actions and experiences through the technologies they create, making design 'the moral activity par excellence' (Verbeek 2011, p. 40). While this is a compelling claim, it risks overemphasising the agency of designers in determining the form of technologies and, in the process, underestimating that of users. More precisely—and contrary to Verbeek's intention—, it can obscure the inherent contestability of technological design and the active role of users in holding designers accountable for undesirable mediations they may inscribe into technological artefacts.

The concepts of (mis)recognition through technology, and specifically design for recognition, are better apt to describe the mediating relationships among designers, technologies, and users. For example, Secomandi and Van Amstel reveal how misrecognition can manifest in user–designer relations through computer-based systems. Building on Hegel's well-known passage on the 'struggle for recognition' between master and slave (or lord and bondsman), they argue that some designed interfaces may coerce people who use them into complying with an alien will—namely, that of designers themselves. In such cases, individuals may internalize the misrecognition imposed on them by others, as if they were merely 'users' of technology rather than as potential designers. They term this process *coerced recognition* and describe it as underpinning a distinct form of social oppression enabled by design (i.e., *userism*), which structurally undermines certain groups' freedom to shape technologies according to their own purposes and to receive appropriate recognition for doing so.

Coerced recognition through designed user interfaces can damage the self-formation of identities at a fundamental level by misrecognizing some humans as *users and only users of computers*. People who could otherwise use technologies to fully develop their human autonomy are contradicted by the specific forms already taken by these artefacts by design. This is especially concerning in contemporary societal contexts where technological design is increasingly inaccessible to diverse groups, even as it plays a growing role in mediating their social relations, both between individuals and between individuals and institutions.

In sum, 'design for recognition' stresses that the design process, the 'material' configurations of technologies, and the effects of technologies influence, and may eventually distort, people's practical relations-to-self and their autonomy, though technology-mediated identity formation. Some potential dangers of this approach are discussed in the next section.

4 Dangers of Designing for Recognition

The move from recognition theory to design practice is not altogether straightforward, in part due to two core elements of Honneth's theory: (a) the relational constitution of identities, which acknowledges that our identities are relationally shaped—including, as we argue here, through technologies, and (b) its phenomenological-empirical starting point, as identifying misrecognition starts from *experiences* of misrecognition, which have a normative potential. Given these two premises, we argue that designing for recognition also entails some dangers, as it may (re)produce instead of battle injustice and unhealthy identity formations. Building on decolonial scholars and critical theorists who discuss recognition as a negative or ambivalent phenomenon (Stahl et al. 2021), we outline four dangers of design for recognition (see also Van Uffelen's forthcoming analysis of the dangers of recognition for environmental justice).

4.1 Moral Intuitions Can Be Distorted

The first danger relates to Honneth's phenomenological starting point, taking collectively shared experiences of misrecognition as a starting point for critique and emancipation. Misrecognition, such as disrespect or devaluation, Honneth assumes, is immediately and intuitively felt as an injustice, as it violates implicit normative expectations. For example, face recognition systems may systematically miscategorise black people due to being trained on racist data, which is immediately felt as misrecognition, leading to mass mobilisation. However, and this is a point that Honneth acknowledges (Honneth 1995, pp. 151–159), one's moral sensorium can already be thoroughly distorted; consequently, a lack of respect, esteem, or love may not always be felt as unjust; instead, it may even feel as deserved. Such distortions may occur when one is socialised in a structurally unjust system, leading to beliefs that one is, in fact, worth less than another. An example is the 'happy housewife', who is disrespected because she is not allowed to work or vote, yet does not experience this as unjust because she has internalised that men are more capable and empowered.

When it comes to design for recognition and its empirical starting point, the problem is clear: if people's moral intuitions are fallible and may be distorted, then designers cannot solely rely on people's experiences of injustice to anticipate, detect, and mitigate misrecognition in technology design. For example, when designing self-driving vehicles, it can become problematic to rely only on people's intuitions as measured through the moral machine experiment (Awad et al. 2018), as people's moral intuitions may be biased. Although Honneth focuses on collectively shared experiences of misrecognition as a better indication of injustice, this does not fully solve the problem at hand, because distortions of moral sensoria might occur on a large scale, and as such, injustices may still remain 'hidden' and reproduced through design (Honneth 1982; van Uffelen and ten Caat 2025).

4.2 Structural Injustices May Be (Re)Produced or Obscured

Second, when designing for recognition, we may (re)produce or obscure other injustices, and this can occur in two ways. First, Nancy Fraser argues that focusing on relations of recognition (understood as cultural depreciation) may distract from, obscure, or give rise to other struggles, particularly those concerning redistribution (Fraser and Honneth 2003). This *problem of displacement* (Fraser 2000) may also materialise through design. For example, to mitigate racial bias in face recognition technologies and allow for proper recognition of a certain user group, developers may train these systems on more elaborate datasets, with more attention to possible stereotypes (Ferrara 2024). However, such efforts often depend on the economic exploitation of workers in the Global South (Muldoon et al. 2024), while also imposing grave sustainability costs on marginalised regions (Strubell et al. 2020). Thus, while designers may enable some users to recognise themselves and avoid oppression in one dimension, they may simultaneously distort identities by positioning those same users as complicit in oppression in another.

Design for recognition may also (re)produce injustice because people's desire for recognition nudges them to accept and conform to dominant social norms, value hierarchies, and identity categories that promise recognition, even though these may be unjust (Allen 2010; Lepold 2021). This can include unjust economic systems, such as capitalism (Althusser 1970), cultural norms imposed by oppressive political systems (Coulthard 2014), or identity categories that are intelligible to others and thus promise recognition, such as the gender binary (Butler 2004). For example, social media platforms may introduce more inclusive emoji or avatar options to improve recognition of users' identities. Yet such design choices may also reinforce the dominance of commercialised, platform-based communication as the primary way of expressing oneself, obscuring alternative modes of interaction that are less driven by profit motives and more supportive of democratic or community-based practices. In other words, design for recognition may fail to critique underlying socio-political assumptions and injustices.

4.3 Value Hierarchies and Social Norms May Become Dominating Once Embedded in Technologies

Third, design for recognition may become problematic and dominating over time. Struggles for recognition aim to alter value hierarchies, social norms, and identity categories, and as such, morality may evolve over time. When designing for recognition, present social relations may become enshrined in technologies. Then, when a normative order becomes outdated, misfits and conflicts may occur between the old and the new. This is not necessarily problematic. After all, technologies inevitably embed value hierarchies and social norms, because they are never value-neutral. The value-ladenness of technologies becomes problematic only when the critique and transformation of the underlying value hierarchies and social norms becomes almost impossible, and ‘opting out’ ceases to be a real option. When technology becomes dominating due to the ‘momentum’ generated by its material structure and embedded values, designers may find it difficult to imagine alternatives, resorting instead to re-inscribing the same outdated identity categories and social norms, corrupting users’ self-relations and leading to misrecognition.

4.4 Identities May Be Reified, Fuelling Political Polarisation

A fourth danger is inspired by what Nancy Fraser described as the *problem of reification.* She argued that struggles for recognition can ‘drastically simplify and reify group identities’ (Fraser 2000, p. 108). Members can be pressured to ‘conform to a given group culture’, leading to a ‘single, drastically simplified group-identity’(Fraser 2000, p. 112) in opposition to other social groups in society. As such, struggles for recognition may easily turn into polarising forces of identity politics, leading to the exacerbation of social conflict. Design for recognition may fall into the same pitfall: when designing for the recognition of certain identities, technologies may end up reifying and ‘fixing’ these identities, paradoxically denying their constitution as complex, dynamic, and intersubjectively shaped. An example can be found in predictive AI technologies that recognize many different identity categories to be as inclusive as possible, such as gender or race, assuming that identities are discrete and static instead of relationally constructed and permanently in flux (see Lu et al. 2022). Although struggles for recognition should always push designers to incorporate a plurality of opposing categories, this opposition must not be construed in terms of self-enclosedness and impenetrability. To move toward a more emancipatory design practice, it is necessary to understand that identities are constituted through mutual, co-determining relationships among subjects within a dynamic totality—a point captured in Hegel’s insight that in true recognition, the ‘I’ is ‘we’ and the ‘we’ is ‘I’. This holistic and dialectical view reaffirms that one’s identity is freely self-confirming only insofar as it both recognises and is recognised by another as equal.

We want to emphasise that these are potential dangers, or risks, not outcomes that necessarily occur when designing for recognition. However, the very nature of Honneth’s recognition theory and its phenomenological and relational starting points makes it possible for us to perceive and understand these dangers.[3] Rather, we consider recognition as an ambivalent phenome-

[3] As such, we do not concur with critics such as Althusser and Lacan who sketch recognition as an intrinsically negative phenomenon that always leads to injustice.

non, as design for (or struggles for) recognition may still have unjust consequences (Laitinen 2021).

5 Conclusions

In this chapter, we highlighted the importance of design in (mis)recognition relations running through technologies. We have argued that technologies, including concrete artefacts and systems, can mediate recognition between humans and shape people's identities in non-neutral ways. Because technologies come into being through design, it follows that identities are, at least partly, shaped by the norms, values, and categories inscribed by others into the material form of technologies. Because design is inevitably a normative activity, there is moral responsibility to devise technologies as configurations of social relations of recognition instead of misrecognition. However, when doing so, designers must be aware that efforts in this direction entail some dangers, including the reproduction and obscuring of injustices and the reification of identities.

Although a full guide for designers is beyond the scope of this chapter, two substantial pointers can be concluded from our conception of 'design for recognition'. First, it urges designers to conceive technologies not as 'anonymous' objects, but as materialised relations of (mis)recognition, inviting a critical gaze. In other words, it focuses the attention of designers on the social relations of giving form to technologies that can adequately support mutual recognition. The proposed unit of analysis, then, becomes relations of (mis) recognition that run through technologies, affecting people's identities, as in their practical relations-to-self and autonomy. Such a perspective has two advantages. On the one hand, when approaching technologies as active mediators of recognitive relations between humans, a critical-emancipatory potential is unlocked for material artefacts and systems: to be involved in a battle against the reproduction of biased structural patterns of identify formation. Technology, when designed to function as a kind of 'translucent mirror', can enable us to see ourselves in it as well as to perceive others gazing upon us through it. This is a crucial step to make us aware of dominant stereotypes and value hierarchies embedded in social life and to question, reject, ridicule, or reimagine pre-existing assumptions.

On the other hand, design for recognition can demystify the agency of nonhumans in human-technology relations by 'bringing back in the loop' other humans, including those who have helped to shape those technologies. At a final stance, it is not purely a technology that misrecognises people;

instead, the source of misrecognition is shared with others entangled in the technology-mediated situation. Such insight can more adequately frame struggles for recognition involving technology, not as between humans and technology itself, but also among social actors who design and use such technology (see Secomandi and van Amstel 2025). For instance, AI systems can be leveraged to take over employee evaluations. Yet, it is not the AI system that recognises the employees, as the AI reproduces the metrics that are most valued by its designers, or the people 'behind the machine' (Anderson 2025). Another example is that of energy models, which are highly influential in energy policy decision-making and are often considered to be purely techno-economic systems. Still, despite their techno-economic appearance, they implicitly contain many normative assumptions, values, and social relations. Given their pervasive impact on energy systems, it is crucial to deconstruct and demystify such seemingly 'neutral' technologies as configurations of (mis) recognition relations, affecting people's relations-to-self, and the design for recognition lens helps deconstruct technologies in these terms. Therefore, the design for recognition perspective is perhaps most valuable in contexts where technologies are usually *not* perceived as configurations of social relations, and AI systems certainly appear to check that box.

Second, design for recognition emphasizes the importance of methods involving stakeholder input and co-design. Technologies can (re)produce undistorted relations of love, respect, and esteem, but they can also misrecognise individuals entangled in those relations as mere users, who cannot and should not contribute to technology design. To elaborate on this point, let us return to the previously mentioned example of care robots, on which elderly individuals are increasingly expected to rely. We have argued that this represents a clear-cut attempt at recognition based on love, according to Honneth's (1995) tripartite model. As it stands, contemporary implementations of such technologies—especially those based on AI—tend to conceptualise the relationship as one between the elderly and a fully autonomous machinic intelligence. If the elderly, in their interactions with these technologies, are further treated according to what Secomandi and van Amstel (2025) call a *userist* design strategy, they are likely to be *coerced* from the outset into self-recognising with the role of a 'user', and nothing more, not as individuals who might also contribute to designing them. In principle, there is no reason why the development of care robots should not create an opportunity for the elderly to embody their own perspectives and care needs in these technologies. But, by experiencing care robots as genuine others, rather than a socially constructed artefact with human originators, elderly users are less likely to resist ageist ideologies embedded in the technology. In that case, the reduction performed by userist design, which oppresses the full humanity of some by hindering their potential to become equal participants in design, is instrumental for the formation of distorted self-relations founded on love.

To conclude, we call for designers to focus their attention on relations of (mis)recognition embedded in technologies, and for philosophers of technology and critical theorists to point their attention to concrete technological artefacts as well as Technology with a capital 'T'. Recognition theory can generate technology critique, and it can point towards avenues of emancipation, much in line with critical theory's ambitions. Such interdisciplinary and empirically-informed efforts would generate a much-needed remedy in a changing world in which technologies increasingly mediate identities in pervasive, and sometimes undesirable, ways.

References

Achterhuis, H., ed. 2001. *American Philosophy of Technology: The Empirical Turn.* Indiana University Press.

Allen, A. 2010. Recognizing Domination: Recognition and Power in Honneth's Critical Theory. *Journal of Power* 3 (1): 21–32. https://doi.org/10.1080/17540291003630338.

Althusser, L. 1970. Ideology and Ideological State Apparatuses. In *Lenin and Philosophy, and Other Essays*, 127–188. New Left Books. http://www.marxists.org/reference/archive/althusser/1970/ideology.htm.

Anderson, J. 2025. Erkenning en Vervreemding op de Digitale Werkplek: AI-Gemedieerde Arbeid. *Wijsgerig Perspectief* 65:16–23. https://doi.org/10.5117/WP2025.2.003.ANDE.

Anderson, J., and A. Honneth. 2009. Autonomy, Vulnerability, Recognition, and Justice. In *Autonomy and the Challenges to Liberalism*, edited by J. Christman and J. Anderson, 127–140. Cambridge University Press.

Awad, E., S. Dsouza, R. Kim, J. Schulz, J. Henrich, A. Shariff, J. F. Bonnefon, and I. Rahwan. 2018. The Moral Machine Experiment. *Nature* 563 (7729): 59–64. https://doi.org/10.1038/s41586-018-0637-6.

Butler, J. 2004. *Undoing Gender*. Routledge.

Coulthard, G. S. 2014. *Red Skin White Masks. Rejecting the Colonial Politics of Recognition*. University of Minnesota Press. https://medium.com/@arifwicaksanaa/pengertian-use-case-a7e576e1b6bf.

Feenberg, A. 2020. Critical Constructivism, Postphenomenology and the Politics of Technology. *Techné: Research in Philosophy and Technology, 24.* https://doi.org/10.5840/techne2020210116

Ferrara, E. 2024. Fairness and Bias in Artificial Intelligence: A Brief Survey of Sources, Impacts, and Mitigation Strategies. *Science* 6 (1): https://doi.org/10.3390/sci6010003.

Fraser, N. 2000. Rethinking Recognition. *New Left Review* 3:107–120.

Fraser, N., and A. Honneth. 2003. In *Redistribution of Recognition? A Political-Philosophical Exchange*, edited by J. Golb, J. Ingram, and C. Wilke. Verso.

Fricker, M. 2007. *Epistemic Injustice Power and the Ethics of Knowing*. Oxford University Press.

Fricker, M. 2018. Epistemic Injustice and Recognition Theory: A New Conversation — Afterword. *Feminist Philosophy Quarterly* 4 (4): https://doi.org/10.5206/fpq/2018.4.6235.

Gertz, N. 2018. Hegel, the Struggle for Recognition, and Robots. *Techné: Research in Philosophy and Technology* 22 (2): 138–157.

Honneth, A. 1982. Moral Consciousness and Class Domination: Some Problems in the Analysis of Hidden Morality. *Praxis International: A Philosophical Journal* 2 (1): 12–25.

Honneth, A. 1995. *The Struggle for Recognition: The Moral Grammar of Social Conflicts*. MIT Press.

Hughes, Thomas P. 1987. The Evolution of Large Technological Systems. In *The Social Construction of Technological Systems*, edited by W. Bijker, T. Hughes, and T. Pinch. MIT Press.

Ihde, D. 1990. *Technology and the Lifeworld: From Garden to Earth*. Indiana University Press.

Jacobs, N., & Huldtgren, A. 2021. Why value sensitive design needs ethical commitments. *Ethics and Information Technology, 23.* https://doi.org/10.1007/s10676-018-9467-3

Laitinen, A. 2021. *On the Ambivalence of Recognition*, 1. Itinerari.

Lepold, K. 2021. How Should We Understand the Ambivalence of Recognition? Revisiting the Link Between Recognition and Subjection in the Works of Althusser and Butler. In *Recognition and Ambivalence*, edited by H. Ikäheimo, K. Lepold, and T. Stahl, 129–159. Columbia University Press.

Lu, C., J. Kay, and K. McKee. 2022. Subverting Machines, Fluctuating Identities: Re-Learning Human Categorization. *ACM International Conference Proceeding Series*1005–1015. https://doi.org/10.1145/3531146.3533161.

Mitcham, C. 1994. *Thinking Through Technology: The Path Between Engineering and Philosophy*. University of Chicago Press.

Muldoon, J., M. Graham, and C. Cant. 2024. *Feeding the Machine*. The Hidden Human Labour Powering AI: Canongate Press.

Olsen, J. K. B., Evan Selinger, and Søren Riis, eds. 2009. *New Waves in Philosophy of Technology*. Palgrave Macmillan.

Secomandi, F., and F. M. C. van Amstel. 2025. Coerced Recognition Through the Service Interface: A Design Ethics Framework for Revealing User Oppression.

Journal of Human-Technology Relations 3:1–25. https://doi.org/10.59490/jhtr.2025.3.7745.
Stahl, T., H. Ikäheimo, and K. Lepold, eds. 2021. *Recognition and Ambivalence.* Columbia University Press.
Strubell, E., A. Ganesh, and A. Mccallum. 2020. Energy and Policy Considerations for Modern Deep Learning Research. In *The Thirty-Fourth AAAI Conference on Artificial Intelligence*. www.aaai.org.
Thompson, M. J., ed. 2017. *The Palgrave Handbook of Critical Theory*. Palgrave Macmillan.
van de Poel, I. 2021. Values and Design. In *Routledge Handbook to Philosophy of Engineering*, edited by D. P. M. N. Doorn, 300–314. Routledge.
van de Poel, I., and P. Kroes. 2014. Can Technology Embody Values? *Philosophy of Engineering and Technology* 17:103–124. https://doi.org/10.1007/978-94-007-7914-3_7.
van Uffelen, N. 2022. Revisiting Recognition in Energy Justice. *Energy Research and Social Science* 92 (August): 102764. https://doi.org/10.1016/j.erss.2022.102764.
van Uffelen, N., and L. M. Santos Ayllón. 2025. Categorizing Experiences of Misrecognition in Energy Contexts: A Recognition Justice Typology. *Applied Energy* 389. https://doi.org/10.1016/j.apenergy.2025.125730.
van Uffelen, N., and S. ten Caat. 2025. Detecting Energy Injustices: Climbing the Ladder of "Hidden Morality". *Energy Policy* 198. https://doi.org/10.1016/j.enpol.2024.114465.
Verbeek, P.-P. 2006. Materializing Morality: Design Ethics and Technological Mediation. *Ethics and Engineering Design* 31 (3): http://www.jstor.org/stable/.
Verbeek, P.-P. 2011. *Moralizing Technology*. Understanding and Designing the Morality of Things: University of Chicago Press.
Waelen, R. 2022. The Struggle for Recognition in the Age of Facial Recognition Technology. *AI and Ethics* 3 (1): 215–222. https://doi.org/10.1007/s43681-022-00146-8.
Waelen, R., and M. Wieczorek. 2022. The Struggle for AI's Recognition: Understanding the Normative Implications of Gender Bias in AI with Honneth's Theory of Recognition. *Philosophy and Technology* 35 (2): 1–17. https://doi.org/10.1007/s13347-022-00548-w.
Winner, L. 2017. Do artifacts Have Politics? *Computer Ethics* May:177–192. https://doi.org/10.4324/9781315259697-21.

Freedom in the Light of Emerging Technologies: Recognition and Hegel's Philosophy of Right

Florian Richter

Abstract Emerging technologies such as artificial intelligence have an impact on our daily lives. Fears about the disruption of political institutions are debated widely. Hegel's political philosophy and his conceptualization of recognition can be used to describe normative standards that help to evaluate developments supported or intensified by emerging technologies. Plato's ideal state is taken as a contrasting conceptualization to Hegel's normative framework. Plato's ideal state, supported by technological means, might turn out to be a technologically-powered dystopia. Freedom as being able to set oneself in relation to the world and oneself is the basis for critical judgment and modern liberal societies. If emerging technologies impede the development of this capacity, they are threatening institutional forms.

Keywords Artificial intelligence • Recognition • Hegel • Human-machine ethics • Philosophy of technology

F. Richter (✉)
School of Transformation and Sustainability, Catholic University of Eichstätt-Ingolstadt, Eichstätt, Germany
e-mail: Florian.richter@ku.de

R. Waelen (ed.), *Normative Relations of Recognition in the Age of AI*, Philosophy and Ethics of AI 2, https://doi.org/10.1007/978-3-032-34028-3_6

1 Introduction

In his *Elements of the Philosophy of Right*, G.W.F. Hegel analyzes practices or institutional forms ("family," "civil community," and "state") which are mediated by different strands of recognition ("love," "particular interests," and "trust"), whereat Hegel understands these forms as realizations of freedom (Hegel 1986, 1991). Emerging technologies, however, pose a risk to these forms since they introduce conditioning factors undermining the possibility of free, symmetric, and reciprocal recognition, which is usually seen as the ideal form of recognitional relations (Brandom 2004). For example, personalized marketing via AI-based systems can undermine the free development of preferences ("particular interests") and can pose a threat by nudging the behavior of consumers. Furthermore, ideas under the titles "smart government" or "smart society" can pose a threat to the "political freedom" of participating in public or governmental affairs. Such developments might also amplify the erosion of trust in and alienation from liberal democratic institutions. Hence, it is of crucial importance to investigate the role of recognition and freedom in the light of emerging technologies. The topic recognition is part of the debate in technology ethics in various ways: (1) Recognition of machines or AI-based systems as (moral) agents is partly the topic of machine ethics; see, for instance, (Moor 2009), and for an overview of machine ethics: (Misselhorn 2022). (2) Others have discussed how humans are (mis-)recognized by technology, see, for instance, (Waelen and Wieczorek 2022). And (3) interpersonal recognition, which is mediated by technology, is partly a topic of postphenomenology (Verbeek 2015; Rosenberger and Verbeek 2015). Although Hegelian ideas about recognition are not explicitly discussed in such approaches, they could be carved out easily, since Hegel and postphenomenological approaches look at how our access to ourselves and to the world is mediated. So, for instance, Hegel's discussion of the practical syllogism in the Science of Logic has been used as a prolific starting point for the philosophy of technology, bringing together mediation and technology (Hubig 2006). Moreover, Gertz, for instance, has used the "mediational theory of Hegel" to understand human-technology relations, particularly the social roles of (military) robots (Gertz 2018). However, Hegel's institutionally-mediated model of recognition has not been discussed in the light of emerging technologies. As I will argue, emerging technologies pose certain threats to freedom. However, Hegel's institutional forms can be taken as confronting such challenges by arguing against, on the one hand, caprice and arbitrariness (despotism) and, on the other hand, conformity. He proposes in his political philosophy social and institutional structures that enable the development of freedom.[1]

The chapter is structured around three key contributions. First, it briefly reconstructs the Hegelian notion of recognition. Second, it takes Plato's *Politeia* (The Republic) as a contrasting conceptualization leading to conformity and a form of equality that stands in opposition to modern liberal states. Third, based on this starting point, Hegel's ideas are described as a normative standard to show how emerging technologies can lead to forms of alienation if not adequately regulated. For that purpose, use cases of emerging technologies and how they pose a threat to the realization of freedom within these institutional forms are discussed.

2 Conceptual Limitations of a Theory of Recognition

I take the institutional forms that Hegel presents ("family," "civic community," and the "state") as a "normative standard," which is probably more in line with the interpretation of Siep (1979) but criticized by Honneth (1995, pp. 189, footnote 2) who aims to elaborate a "theory of recognition" that can be understood as describing "necessary conditions for human socialization." Nevertheless, it seems also that Honneth believes that recognition and socialization are institutionally mediated, i.e., they come with ascription of roles and entitlements that cannot solely be instituted by interactions between individuals. However, he takes the "experience of disrespect" as the basis of "social conflicts" (genesis of recognition) from which positive relations of recognition spring. They serve then as a normative standard (Honneth 1995, p. 118). Since I will not propose a theory of recognition that is grounded in a particular normative principle, such issues can be disregarded. Nevertheless, the limitations of a theory, as Honneth envisions it, should be discussed shortly. Such a theory of recognition is not feasible because Honneth would not only need to determine the "necessary conditions for human socialization" but also the sufficient condition that initiates the transition from social conflict to recognition. Hence, assuming that from social conflict recognition necessarily follows means to be able to determine the necessary and sufficient conditions. I believe that Honneth does not commit himself to such a strong

[1] Hegel has often been taken as the defender of the Prussian state and monarchy. My aim is to distill the main ideas of Hegel where he develops a conceptualization of a modern liberal state that contrasts clearly with Plato's ideal state. Aspects where he supports a monarchy are therefore neglected.

claim since he wants to develop a more empirical approach to recognition. However, in order to develop an empirically-based materialistic theory of recognition, one must show how, from social conflicts, forms of recognition necessarily follow. Even if it were possible to test certain hypotheses under laboratory conditions, it would not be possible to determine whether this would apply on the societal level.

Honneth takes the early writings of Hegel as a starting point. However, in his later writings, Hegel does not start with social conflicts or a state of nature (like Thomas Hobbes). He also does not believe that building a society leads to a kind of alienation from the state of nature (Jean-Jacques Rousseau). The paragraphs about the "objektive Geist" start with the "right" ("Recht"), where Hegel discusses the concept of "person." Subjects as persons can own property and make contracts. He even criticizes such social contract theories that use the state of nature as a starting point, since they are based solely on violence and caprice (Hegel 1991, pp. 392–397, §§ 488–502) Persons can freely and willfully commit themselves to obey contracts that they made with other persons. In the next section, Hegel determines the moral subject, who is able to critically judge and evaluate cases and situations within certain contexts (Hegel 1991, pp. 397–402, §§ 503–512). All of these conceptual developments must be kept in mind when looking at the institutional forms, since within those forms the subject is able to act freely, judge critically, evaluate cases and the conditions of such institutional forms. Hence, institutional forms where this is not possible can be judged as not morally desirable and exhibiting forms of alienation. Here, I will take a look at how technological conditions might make it impossible for subjects to develop such capabilities to judge critically or hinder them from reflecting on their preferences, etc.

Alienation presupposes that someone is alienated from a certain state, i.e., an individual is not her true self or has not reached her true consciousness. It is the role of the philosopher who is able to think deeply to show the individual the enlightened path to a just society (Plato or Karl Marx) or to a life with existential peaks (Martin Heidegger in *Being and Time*). However, this is a problematic notion because it is assumed that the philosopher knows either the ideal form of the state, society, or human existence. Plato's *Politeia* serves as a paradigmatic exemplar here because he introduces the philosopher king as someone who has a deeper understanding, which makes him or her best suited to govern. While Plato takes equality and unity as a feature of his ideal state, Hegel's philosophy of right steers toward inequality as a feature of the modern state. Hegel is, as I understand him, a more humble thinker who assumes that we must be capable of setting ourselves in relation to ourselves and others, the capability to recognize oneself and others. I will argue that this makes Hegel's philosophy better suited to deal with illiberal tendencies and dogmatisms that can spring from emerging technologies.

3 The Civic Community

I will describe Plato's ideas from his dialogue *Politeia* and use it as a contrasting conceptualization of Hegel's political philosophy. Plato's ideas serve as a philosophical elaboration of what might be realizable via emerging technologies.

3.1 Plato: Equality and Unity

In his attempt to describe the ideal state, Plato introduces features of equality that have been discussed widely in the history of philosophy. According to him, the guardians should not have private property (neither gold nor immovable property). It would make the guardians, who should rule over the city-state, anxious that they could lose their fortune and, in the end, have an interest in the mundane businesses, which would make them bad rulers because they should be only interested in living up to the just order, since acquiring gold or other things would make them like to be recognized as superior by others ("megalothymia"). Instead, they should aim for a kind of recognition as equals (at least within the class of the guardians) (Plato 1991, 415d-417b).[2]

Another feature of the ideal state is a kind of uniformity in what is considered to bring pleasure and pain. Plato compares the state to the body, and when one part is hurt, like the finger, the whole body feels it. Similarly, the whole community should feel when one part is hurt. (461e–466d) Such compassion could also count as a form of recognition. Nevertheless, having just compassion with someone might not be enough. It is also important to have standardized measures to help the person. Who would like to go to a medical doctor who feels with you but does not prescribe any medicine to relieve the pain?

Both features of the state are aspects of unity, which is the overall goal of the state. Plato thinks that only a strong sense of unity can hold together the state. Aristotle, on the other hand, thinks a state needs plurality; otherwise, it

[2] It seems to me that the more radical notion that also the farmers and handymen should have no private property is not made by Plato because he thinks that they, in a certain way, do not share the noble nature of the guardians who actually could handle a life being disinterested in mundane businesses.

will collapse. (Aristoteles 1995) (1261a) Plato wants to suppress personal interests and motivations in their plurality that diverge from the idealized unity. In a nutshell, the differences in what brings one pleasure and the other not should be eradicated so that we all feel pleasure for the same thing. This is also achieved through censorship of poetry and literature, where gods and heroes are portrayed as evil, in order to form the soul as good. (381a-394c) Plato's notion of education has the advantage that it promotes a kind of social mobility where someone from the working class with the right talents can also become a guardian. While merit is thus connected to educational excellence and not based on something arbitrary like birth in an aristocracy, it still dis-values forms of life and gainful occupation. Although it is not completely our choice, what we value, what brings us pleasure, what we would like to own, and which occupation we consider adequate for us, we would hardly like to give up such options of choice, and that they are made by others for us.

3.2 Hegel: The System of Need and Preferences

3.2.1 System of Needs: Market

Hegel's "system of needs" is characterized by features that can also be found in Adam Smith, for instance, the "division of labor" and the "production of exchangeable means via one's own work." It is a "mediation of satisfaction via the work of All." He distinguishes the system of needs in three sectors or estates: the first estate encompasses agriculture, which provides food, the second estate is mainly concerned with trade and production, and the third estate is in charge of public administration. Hegel characterizes the last one as the "thinking estate" because it is concerned with the "general interests" ("allgemeinen Interessen"). So, it seems like this one is higher in esteem and similar to Plato's guardians. Hegel proclaims that individuals are part of each estate based on "natural talent, skillfulness, arbitrariness, and luck." (Hegel 1991, p. 406–407, §§ 524–528) In a way, it seems that Hegel believes that individuals can develop their talents to move from one estate to another and that there are contingencies of life involved that might constrain also choice options. Also, this point seems similar to Plato's idea of educational excellence and social mobility.

For participating in the systems of needs, one needs a "formal education" ("formelle Bildung"), and what Hegel writes about the second estate seems also to be true about the others: the individual needs "subjective skill, talent, reason, and effort" to succeed. (Hegel 1991, p. 406–407, §§ 524–528) Formal education enables one to participate in the free market, land a job, and gain money. However, Hegel makes a big difference here from Plato, since all three estates are in a way involved with their interests in the world because, through their skills, they offer their labor on the market. Even the public officers who work for the state do so in some way since they are paid for their service.

The idea of talents and skills and the separation of work also leads to breaking down tasks into smaller tasks, so that "skillfulness becomes mechanical and enables that human work can be substituted by the machine." Hence, formal education is something different from a broader kind of formation, which is described below as a presupposition to participate in public affairs. Nevertheless, by belonging to one of the three estates, the individuals have their "recognition and honor." (Hegel 1991, pp. 405–407, §§ 523–528) Honor might be usually used in contexts of fighting wars. However, Hegel uses it for being recognized for one's own work, talents, and skills in the market to provide for oneself and others. Being recognized for the work one does and being needed based on one's talents and skills is important, but has also broader implications. Automation in factories led to job losses in the twentieth century, and one of the fears regarding AI is that it might automate so-called white-collar jobs, leading to job losses in that sector.[3] Massive job losses +and high rates of unemployment exclude people from being recognized this way. Despite the economic downsides, it might also have implications for people's well-being. Being needed and being able to provide for oneself are essential features of our self-understanding to pursue goals.

How can the system of needs prevent not only the loss of jobs but also help the individuals in the system? The "blind necessity of the system of needs" does not provide in a systematic way for a kind of "social welfare," only maybe via charity or philanthropy. The "purpose" of the "civic community" is the "safeguarding of the satisfaction" of the needs. However, since needs change, "fallacies and deception" can disrupt parts of the "machinery," and the persons only have a "conditioned ability" to gain money, a criterion of adequacy must be established to take the individual as a "morally qualified purpose." Conditioned ability means that the talents and money via inheritance are unequally distributed, and the civic community cannot "safeguard the satisfaction of the individual." That is why the state is needed to establish institutional structures that protect the individual's development. However, also via

[3] The rise of right-wing political parties and movements in Western democracies might be connected with the loss of jobs, and, hence, the denial of recognition of blue-collar workers.

"corporations," the individual might be able to safeguard her "assets." (Hegel 1991, pp. 411–412, §§ 532–534).

3.2.2 Preferences

Hegel describes in the chapter "Lordship and Bondage" ("Herr und Knecht") in his Phenomenology of Spirit, not a struggle between two different subjects but a struggle within one subject. The most common reading has been the former one. However, as some have proposed (Hubig 1985; Stekeler-Weithofer 2008; McDowell 2009), the latter makes more sense, since Hegel describes modes of appearance of consciousness, and he starts discussing interactions in the later chapter "Spirit." Hence, it would be a strong inconsistency on the side of Hegel to introduce a struggle between subjects at this point in the text.

Animals have desires. They search for food and devour it. Hegel writes that "work is, however, constrained desire." The master (mind) mediates the access to the world via the slave (body), who "is immersed in the diffusion of life" and has to deal with the independence of things. Hence, the master recognizes the slave for doing the work, while the master "enjoys in a pure way." Humans distance themselves from the immersion in the diffusion of life by working with things that are independent and serve as means to achieve certain desires. However, the world has a resistance, or it constrains desire. While animals feel pleasure immediately, humans can mediate pleasure to sublimate it. This means that humans can set themselves in relation to their preferences and the pleasure they expect from them. Hegel calls this kind of freedom "obstinacy" ("Eigensinn") and the kind of formation "skillfulness." (Hegel 2008, pp. 137–155).

The desire to cook a meal is constrained by the availability of the food, its quality, and my skill to prepare it. The skill can only be formed through experience and by being immersed in life. I can also set myself in relation to the product and how it is in comparison to how I desired it to be. The experience of this difference is essential to improve my skillfulness in cooking. Moreover, the formation of skills enables me to evaluate the products and skills of others. In a highly diversified economy with many different tasks and products, the medium money enables comparability. Maybe I am not able to judge the skillfulness to perform a certain task; however, I might be willing to spend a portion of my salary to buy the product someone offers. So, via the development of talents and skills, we develop different interests and do not form a conform idea of what brings pleasure and pain.

3.3 Alienation Via Emerging Technologies

Emerging technologies, like AI, might now have an impact on the one hand on being recognized for one's talent and skillfulness to perform well on the labour market, since it might lead to disruption in, for instance, white-collar jobs. On the other hand, new technologies, like AI-powered neurotechnologies, might make it impossible to form preferences based on acquired skills. What Plato described as a philosophical utopia might turn out to be a technologically-powered dystopia.

3.3.1 Formation of Skills

To mitigate the impact of AI-based technologies on the labour market, it has been proposed to introduce a universal basic income. For that purpose, people could reorient themselves and prepare themselves for new challenges in the job market. Several advantages of a universal basic income can be mentioned, such as that it eliminates the need for government programs, supports caregivers, reduces income inequality, etc. Also, disadvantages can be listed, such as that it might reduce the incentive to work, comes with high costs, and gives people money who do not really need it. However, the recognition and honor that Hegel describes when one is able to gain income based on talents and skillfulness might be reduced since the income is unconditional and thus not based on the "conditional ability." Talents and skillfulness are accidental (zufällig), and hence, conditional, i.e., based on inheritance, place of birth, opportunity to get a certain education, etc. For instance, someone who is talented enough to become a football player and has the opportunity to develop such a talent might be able to earn a lot of money. Centuries ago, such a talent would not have been helpful to earn money. Although we are dependent on many conditions, we are also proud of how we develop our talents and skills to perform in the free market. In times of personal failure to provide for oneself, it is wrong to blame this person for having developed the wrong skills, because of the accidental nature of skills, talents, etc. However, in times of personal success, we should find recognition in having developed the right skills without fancying oneself too much. Yet, recognition is expressed mainly in monetary compensation (although companies also provide additional alternative forms of compensation). The problematic issue of a universal basic income might not be so much that it takes away the incentives to work, but that it might not be possible to receive recognition for the skillful work one is able to perform.

Even on a more extreme account, as Plato suggests for the cast of the guardians, where private property is abolished, the recognition of one's own work will also disappear. Nevertheless, one might argue that other forms of recognition might be possible that will, in the end, even be more fulfilling. Such a society might also be envisioned on technological grounds as a kind of fully automated communism. However, it is questionable whether something like this would be viable or even desirable.

3.3.2 Formation of Preferences

Marketing is a business activity researching the needs and preferences of customers so that products can be developed that meet their needs and preferences. Nevertheless, this can also lead to manufacturing customers' needs where mere influence in advertising is converted into manipulating customers to buy into unwanted needs. Newer developments in the field of neurotechnology make it possible to investigate the needs of customers. For instance, non-invasive methods, such as electroencephalography, could be used to investigate neural stimulations when watching advertisements or tasting food products. Fields emerged, such as "consumer neuroscience," "neuroeconomics," and "neuromarketing." Somehow, the promise of such technologies is that the real consumer needs or preferences can be revealed or validated directly from the brain, instead of conducting behavioral experiments or doing questionnaires. The results could be used to personalize taste experiences, advertisements, designs, etc., to a point where consumers cannot resist buying such products. Although noninvasive technologies are used, one could argue that it is an intrusion into the mental privacy of the consumer. Even though there has been extensive research on neuromarketing in the field of business research (Bhardwaj et al. 2023; 2024) or in neurosciences (Mansor and Mohd Isa 2020), ethical investigations are still scarce.

It is wrong to deceive someone when trying to sell a product or to use dark patterns (Gray et al. 2021; Nie et al. 2024) to trick someone into clicking and buying a product. These are issues of general moral concern applied to concrete cases and can be handled by general moral standards. Hence, we do not need specific technology or business ethics for such issues. Instead, it should be investigated how certain neurotechnologies enable deception, manipulation, nudging, and the undermining of human autonomy (Hubig 2007, pp. 16, 169). Neurotechnologies to reveal customer needs seem to be on another level since they bypass how needs and preferences are developed. For Hegel, the development of preferences is connected to the formation of skills and being able to value what is produced. Investigating what triggers consumers' brains to buy a product, or which kind of products they prefer, seems to produce a kind of uniformity in what brings everyone pleasure without considering the specific talents, skills, and character traits. For instance, for developing a taste for movies, books, music, paintings, etc., one must do the irksome work to acquire skills like writing or playing an instrument. Even on a basic level to develop a taste for movies, which might be acquired by almost pure consumption of them, one must also watch movies one does not like. Movie recommendation systems that just present someone with what they like will lead to uniformity: the person likes comedies of a certain type and gets recommended comedies of a certain type. Moreover, not being aware anymore of why I chose a certain product or which kind of advertisement techniques are used to influence me not only undermines human autonomy but also the idea of a free market where sellers compete with each other, offering the "best" product because they expect that it will be bought.

4 The State

In Plato's state, there is equality among the guardians. Political participation is only possible if one is blessed with good talents to get into the caste of the guardians. Moreover, he assumes that a kind of conformity of opinions can be reached through censorship. Hegel, on the other hand, believes that a kind of plurality among opinions, talents, and skills serves the state and should be fostered via education.

4.1 Plato: Selection of the Guardians and Censorship

The selection of the guardians and the ones who rule in the city-state is based on their natural talent and performance in education. Instead of an aristocratic elite where status is inherited, Plato proposes a meritocratic nobility where excellence must be shown in the educational training as a philosopher (Plato 1991) (535a–541b). This means also that the persons are not free to choose their profession or life plan. From a modern perspective, this is paternalistic by pushing people to play a certain role in society. Nonetheless, Plato's picture is in accordance with the Ancient notion of fulfilling a certain role in society, and he also limits participation in public affairs to a small elite of philosophers; similar to the political participation in Greek city-states, where only a small number of people were allowed to govern.

Another problematic aspect of Plato's ideal state is his notion of reproduction within the guardian class. He sketches a programme that is similar to animal breeding, where the best ones are selected to procreate. In order not to create upheaval, the selection process should be disguised as a lottery. Moreover, the parents should at the end not know who their children are. In the working class, on the other hand, people would still live in traditional families. (457c-461b).

One more aspect that has been mentioned is the idea of censorship, where poetry or literature serves to form the soul in a good and harmonious way by not portraying any evil. (381a-394c) Also, here, the elite class selects adequate media that are useful for education. The students should not see content that might lead them to do evil things. Such parental control also exists for children nowadays. However, Plato's idea is much more restrictive because even adults should not see such content. Plato's ideal state resembles a kind of paternalism where a small group decides what is good for others and what kind of behavior is adequate. And, he opts for a programme of population control similar to animal breeding.

4.2 Hegel: Family and Education

4.2.1 Family

The first institutional form in the Sittlichkeit is, for Hegel, the family. It is based on the reciprocal recognition of "love." Of course, Hegel bases his idea on the "difference of the natural sexes." However, an important point for him is the "second birth" of the children, which is "education" ("Erziehung") (Hegel 1991, pp. 404–405, §§ 518–522). In a modern liberal state, other forms of having children (from adoption to surrogacy) are possible, and same-sex relationships that have children and form a family. The bounding factor is always "love." Plato, on the other hand, wants to establish for the guardians a system that is not based on love and to sublimate the relationship between, on the one hand, partners and, on the other hand, between parents and children to a kind of rationalized form. Hegel calls the family "empfindender Geist" (Hegel 1991, p. 404, § 518). The family as a loving nucleus of society is less controllable and steerable than state-run programmes of population control, which can also be technologically enhanced. The idea is that the partner is chosen freely and based on love and not because it serves the community, the political agenda of the family, etc.

4.2.2 Education

Freedom is the "security of property, as the possibility to develop talents and good characteristics." The state guarantees that the citizens can develop their talents in accordance with their particular interests and what brings them pleasure. They should build a kind of "moral independence" to judge critically and be able to reflect. Subjective freedom can only grow under the "conditions of objective freedom" in modern states. "Through formation" ("Bildung"), "humans make themselves more unequal." (Hegel 1991, p. 416, § 539) Hegel does not aim for conformity in opinion and lifestyles but for the free development of persons. While Plato's notion of education is centered around a conformity where certain ideas, opinions, or expressions are suppressed by a small group, Hegel believes that differences are valuable and should be embraced. The state should guarantee that individuals can develop their talents, and such kinds of societies will realize freedom better by shaping their educational institutions to produce individuals with unequal skills.

4.3 Alienation Via Emerging Technologies II

4.3.1 Genetic Engineering

Technology as a medium is seen as an "extension" of the human (McLuhan 2013), and at one point, there might be a tight coupling of "human brains and computing machines" (Licklider 1960). Such ideas were decades ago, rather science fiction, but have become reality in brain-computer interfaces and are mostly present in the so-called movement "transhumanism" where technological enhancement of humans is sought. This enhancement is discussed and envisioned based on existing technologies, but can also be more speculative and ranges "from regenerative medicine to nanotechnology, radical life extension, mind uploading and cryonics, among other fields. Distinctive currents coexist in transhumanism, such as: libertarian transhumanism, democratic transhumanism, and extropianism." (Ferrando 2013, p. 27) Transhumanism emphasizes "notions such as rationality, progress and optimism" (Ferrando 2013, p. 27). Some have also discussed cognitive enhancement via genetic engineering to make humans more intelligent (Shulman and Bostrom 2014). The concept "intelligence" is extremely difficult to determine and operationalize, and very limited determined when the only measurement is via standardized IQ tests, as Shulman and Bostrom intend. Two paths seem here possible: either cognitive enhancement is market-driven, where some intend to make their offspring more intelligent (while still more or less regulated by states), or the development is state-owned, where programs are employed to increase the IQ of the people.

Similarly, to Plato, it should be a selection of the more intelligent or apt ones for breeding. Aristotle already criticized Plato's ideas, which would disrupt the family as a foundational structure of society. Among his arguments against Plato, he brings forward that it will lead to a disembedding where no one will feel directly responsible for raising the children since they belong to everyone; a kind of responsibility diffusion. However, this notion is somehow problematic since in ancient Greece, women and children were the property of men. Hence, the care that Aristoteles describes is more grounded in self-interest for property. Nevertheless, he mentions another aspect that is more in accordance with Hegel's idea of reciprocal recognition in the form of love. Aristoteles criticizes Plato's idea of a community since it will lead to the dilution of friendship among the citizens. Forming a friendship between parent and child is essential because the relationship is then grounded in love. The family is a place where unconditional love is first experienced and will also lead to trusting relations among citizens (Aristoteles 1995) (1260b-1262b).

The genetic engineering of offspring, either sought by parents or the state, is, in a way, understanding children as property that can be instrumentalized for a certain cause. Although reciprocal recognition in the form of love among family members is the ideal, in families, forms of instrumentalization and manipulation also occur. Nevertheless, it should be clear that genetic engineering, particularly on societal levels via state-owned programs, and beyond preventing awful hereditary diseases, is not morally desirable.

4.3.2 Education, Censorship, and Social Credits

Another form of alienation would be through instrumentalizing citizens by censorship and by providing education that leads to conformity rather than to "inequality," as Hegel argues. This also entails that forms of "dogmatic upbringing and oppression" that "undermine the autonomy of decisions" must be avoided. Hubig calls this dimension "legacy values." (Hubig 2007, p. 142) It must be ensured "that the subjects can develop as autonomous subjects of evaluation. This includes maintaining the social structures that first and foremost allow the formation of personal identity." (Hubig 2007, p. 142) However, this is not a conservatism of traditional structures; rather, social structures should be established to enable individuals to set themselves in relation to their options, wishes, values, and preferences, and, if they need to, can protest against these structures (Hubig 2007, p. 142). Consequently, forms of protest must also be possible that can be sought without (violent) punishment, so that the individuals are enabled to think critically, understand their needs, and convert them into achievements of well-being without ideological impositions.

Otherwise, formation would only be a kind of ideological conditioning that is achieved through censorship in education or by sanctioning morally good or bad behavior, at least, what one group might consider morally desirable. Something like this is encapsulated in the social credit system in China, where certain behaviors are listed on a so-called Redlist and considered morally good, and other behaviors are considered morally wrong and listed on a so-called Blacklist. Good behavior is rewarded, and bad behavior is sanctioned (Engelmann et al. 2019). Moreover, the social credit system is supported by AI-based systems, like facial recognition, to make it more efficient and effective. Hence, a social control mechanism, which is in education done by censorship, as in Plato's ideal state, is extended into the social life to condition the citizens' behavior. Several problematic points can be mentioned: First, the citizens are not included in publicly debating about what counts as morally good behavior. Second, conditioning behavior undermines the autonomy of the citizens and makes it impossible for them to set themselves in relation to their behavior and choice options without coercion or the threat of punishment. Third, it brings conformity to behavior, ways of life, and opinions that undermine the pluralistic aspect of modern liberal societies.

According to Hegel, laws shall constrain the caprice and particular interest of individuals. However, they are based on their "free will" and "ethical attitude" ("Gesinnung"). That the laws and the constitution are "reasonable" ("vernünftig") lies in the responsibility of the people (Hegel 1991, pp. 414–417, §§ 538–540). Hegel's (in)famous dictum from the prologue of his philosophy of rights—"Was vernünftig ist, das ist wirklich; und was wirklich ist, das ist vernünftig." (Hegel 1986, p. 24)—should not be read as a statement about how things really are but in light of how the laws inhibit arbitrary treatment and how they reflect the free will of the individuals. In this context, Hegel criticizes the idea that in front of the laws everyone is equal. While he states that this contains a deep truth, it is a mere tautology. On the one hand, it presupposes that humans are equal and not unequal in status, as in Rome or Greece, where only some were considered equal and were allowed to rule. On the other hand, it is problematic because, in front of the court, unequal treatment is necessary when it is well justified. There are arbitrary differences in character, age, circumstances, etc. that also justify unequal treatment when it comes to crimes (Hegel 1991, pp. 415, § 539). For instance, someone with a mental disorder might not go to jail for a crime but to a psychiatric clinic. Such treatment is far from making citizens conforming creatures according to pre-established "moral" codes.

5 Freedom and Alienation: Hegel's Path to Inequality

Plato thought that a kind of uniformity of pleasure, preferences, and moral behavior leads to a harmonized and just society. He wanted to achieve this via censorship, education, and the selection of individuals to put them in different social classes. Emerging technologies could enable and enhance such an idealized state via neuromarketing, genetic engineering, surveillance, and control. However, this would undermine modern liberal societies and lead to different forms of alienation. For instance, not being able to set oneself in relation to preferences and develop them independently, is not only problematic from an individual perspective, but also on a systemic level, since it undermines free markets. Genetic engineering could undermine the institutional form of "family" and its form of recognition (love) by instrumentalizing offspring and, thus, alienating parents and their children. Surveillance and control might lead to alienation by forcing people to act in accordance with a certain established "morally good" behavior. This kind of conformity makes it impossible to develop critical thinking since also protest, deviation from standards, etc., are necessary to develop such an essential ability. Hegel's institutionally mediated recognition serves here as a guidepost to understand the problematic issues that emerge with new technologies.

So, in the end, Hegel argues that modern liberal states should treat their citizens unequally and encourage them to develop their talents and skills in different ways. This will make society more resilient and better prepared for disruptions. The state also has the task of enabling the active development of different talents and skills, which at one point might not seem useful for society. For example, some decades ago, sinology was considered a so-called "small subject" to study. Nevertheless, the world changes, and to really grasp and evaluate the impact of China in the economic, cultural, and political spheres, researchers are needed. Fostering inequality in this sense makes societies better prepared for disruptions, not only on economic, political, and cultural grounds but also on technological grounds. At least, fostering education and (ethical) research about emerging technologies might enable us to reflect on their implications for us and society.

References

Aristoteles. 1995. *Politik*. Hamburg: Meiner.

Bhardwaj, S., G. A. Rana, A. Behl, and S. J. Gallego de Caceres. 2023. Exploring the boundaries of Neuromarketing through systematic investigation. *Journal of Business Research* 154.

Bhardwaj, S., S. Thapa, and A. Gandhi. 2024. Advances in neuromarketing and improved understanding of consumer behaviour: Analysing tool possibilities and research trends. *Cogent Business & Management* 11 (1):

Brandom, R. 2004. Selbstbewusstsein und Selbst-Konstitution: Die Struktur von Wünschen und Anerkennung. In *Hegels Erbe*, edited by C. Halbig, M. Quante, and L. Siep, 46–77. Frankfurt am Main: Suhrkamp.

Engelmann, S., et al. 2019. Clear sanctions, vague rewards: How China's social credit system currently defines 'good' and 'bad' behavior. FAT* 2019 - proceedings of the 2019 conference on fairness. *Accountability, and Transparency*69–78.

Ferrando, F. 2013. Posthumanism, Transhumanism, Antihumanism, Metahumanism, and new materialisms: Differences and relations. *Existenz* 8 (2): 26–32.

Gertz, N. 2018. Hegel, the struggle for recognition, and robots. *Techné: Research in Philosophy & Technology* 22 (2): 138–157.

Gray, C. M., J. Chen, S. S. Chivukula, and L. Qu. 2021. End user accounts of dark patterns as felt manipulation. *Proceedings of the ACM on Human-Computer Interaction* 5 (CSCW2): 1–25.

Hegel, G. W. F. 1986. *Grundlinien der Philosophie des Rechts*. Frankfurt am Main: Suhrkamp.

Hegel, G. W. F. 1991. *Enzyklopädie der philosophischen Wissenschaften (1830)*. Hamburg: Meiner.

Hegel, G. F. W. 2008. *Phänomenologie des Geistes*. Frankfurt am Main: Suhrkamp.

Honneth, A. 1995. *The struggle for recognition: The moral grammar of social conflicts*. Cambridge: Polity Press.

Hubig, C. 1985. *Handlung—Identität—Verstehen: Von der Handlungstheorie zur Geisteswissenschaft*. Weinheim/Basel: s.n.

Hubig, C. 2006. *Die Kunst des Möglichen I: Technikphilosophie als Reflexion der Medialität*. Bielefeld: Transcript.

Hubig, C. 2007. *Die Kunst des Möglichen II: Ethik der Technik als provisorische Moral*. Bielefeld: Transcript.

Licklider, J. 1960. Man-computer Symbiosis. *IRE Transactions on Human Factors in Electronics* 1:4–11.

Mansor, A., and S. Mohd Isa. 2020. Fundamentals of neuromarketing: What is it all about? *Neuroscience Research Notes* 3 (4): 22–28.

McDowell, J. 2009. The Apperceptive I and the empirical self: Towards a heterodox Reading of 'Lordshipand bondage' in Hegel's phenomenology. In *Having the world in view, essays on Kant, Hegel and Sellars*, 147–165. Cambridge: s.n.

McLuhan, M. 2013. *Understanding media: The extensions of man*. Berkeley: s.n.
Misselhorn, C. 2022. *Grundfragen der Maschinenethik*. Stuttgart: s.n.
Moor, J. 2009. Four kinds of ethical robots. *Philosophy Now* 72:12–14.
Nie, L., et al. 2024. Shadows in the Interface: A comprehensive study on dark patterns. *Proc. ACM Softw. Eng.* 1 (FSE): 204–225.
Plato. 1991. *Politeia*. Frankfurt am Main/Leipzig: Insel.
Rosenberger, R., and P.-P. Verbeek. 2015. A Postphenomenological field guide. In *Postphenomenological investigations: Essays on human-technology relations*, edited by R. Rosenberger and P. Verbeek, 9–41. New York/London: s.n.
Shulman, C., and N. Bostrom. 2014. Embryo selection for cognitive enhancement: Curiosity or game-changer? *Global Policy* 5 (1): 85–92.
Siep, L. 1979. *Anerkennung als Prinzip der praktischen Philosophie: Untersuchungen zu Hegels Jenaer Philosophie des Geistes*. Freiburg/München: Alber.
Stekeler-Weithofer, P. 2008. Wer ist der Herr, wer ist der Knecht? Der Kampf zwischen Denken und Handeln als Grundform jedes Selbstbewußtseins. In *Hegels Phänomenologie des Geistes: Ein kooperativer Kommentar zu einem Schlüsselwerk der Moderne*, edited by K. Vieweg and W. Welsch, 205–237. Frankfurt am Main: Suhrkamp.
Verbeek, P.-P. 2015. Beyond interaction: A short introduction to mediation theory. *Interactions* 22 (3): 26–31.
Waelen, R., and M. Wieczorek. 2022. The struggle for AI's recognition: Understanding the normative implications of gender bias in AI with Honneth's theory of recognition. *Philosophy & Technology* 35 (2): Article 48.

What's Love Got To Do With It? Children, Recognition, and Algorithmic Technologies

Michał Wieczorek

Abstract This chapter offers a critical assessment of Axel Honneth's recognition theory by focusing on children's experiences of (mis)recognition. Building on Honneth's discussion of primary relationships of care (i.e., love), I argue that the asymmetry and vulnerability characteristic of such relationships are also present in children's struggles at the level of rights and solidarity. Recognition of children needs to acknowledge that children think and feel more than they can name and express. However, this requires good will and interpretative work on the part of adults who can support children under their care in articulation of recognitional claim or overrule their input (e.g., due to paternalism or abuse of power). Drawing on social work and child care literature, I introduce three characteristics a recognising subject (e.g., an adult) needs to possess to successfully recognise children: (1) ability to develop meaningful and in-depth relationship with children (i.e., familiarity with a particular child and their experience), (2) hermeneutic capacities (i.e., ability to make sense of indirect and non-verbal recognitional claims) and (3) ability to make prudential judgments (i.e., determining which aspects of the interaction are significant for the struggle for recognition). This development of recognition theory allows me to consider the impact of algorithmic technologies on children's self-development. Such tools and services increasingly form a

M. Wieczorek (✉)
School of Education, University College Dublin, Dublin, Ireland
e-mail: michal.wieczorek@ucd.ie

R. Waelen (ed.), *Normative Relations of Recognition in the Age of AI*, Philosophy and Ethics of AI 2, https://doi.org/10.1007/978-3-032-34028-3_7

part of childhood and mediate children's interactions with others and their environment, but existing work on recognition and technology only examined the experience of adults. I argue that algorithmic technologies do not possess the characteristics required to successfully and appropriately recognise children and are unlikely to simulate them. I conclude by highlighting areas where children's interactions with technologies are particularly likely to have constitutive impacts and which warrant further attention from recognition scholars.

Keywords Recognition • Children • Technology • Love • Vulnerability • Axel Honneth

1 Introduction

This chapter critically engages with Axel Honneth's recognition theory, as outlined in his seminal *The Struggle for Recognition* (Honneth 1996), to interrogate how children's role in recognitional struggles is and should be conceptualised. Honneth draws on Winnicot's (2018) work on primary care-giving relationships between mothers and children and Mead's (2015) theory of social development to introduce the first of his relations of recognition—love. However, even though Honneth wants to develop a theory of both social and individual development, children are largely absent in his overall writings. Honneth's analyses often detail struggles for recognition between adults as he discusses the liberation struggles undertaken by women, people of colour, or members of the working class. The omission of children is particularly striking if we consider that as not yet fully developed subjects, they are both in particular need of positive recognition of their needs, capacities and accomplishments, and especially vulnerable to experiences of misrecognition. And while Honneth did offer some brief discussion of children in his later works (Honneth 2012, 2014), critics accuse his theory of being adult-centric—treating children as "adults in waiting" (Thomas 2012) and failing to consider the unique and complex nature of their struggles for recognition (see also Thomas et al. 2016; Warming 2015).

For this reason, I draw on social work and child care literature which in recent years has increasingly used Honneth's recognition theory to problematise children's experiences of recognition or misrecognition, e.g., in the context of foster care or political participation (Mitchell 2022; Paulsen and Thomas 2018; Thomas 2012; Thomas et al. 2016; Warming 2006, 2015), and to propose normatively informed frameworks for engaging with children and recognising their experiences (Häkli et al. 2018; Houston and Dolan 2008; Smith et al. 2017; Warming 2006). This allows me to advance philosophical work on recognition theory in two interrelated ways. First, I outline the differences between children and adults as actors in the struggles for recognition and demonstrate where and how recognition theory needs to be further development if we want it to account for the specificity of children's experiences of (mis)recognition. I demonstrate that in their attempts to have their needs, capacities and accomplishments recognised, children are more vulnerable to paternalism and power differences, but they also depend on the support of adults who often have to engage in interpretative work (e.g., to make sense of children's recognitional claims) and exercise a great deal of practical wisdom (e.g., to balance children's right to self-determination with the responsibility to shelter them from consequences of bad decisions). For this reason, I introduce three characteristics a recognising subject (e.g., an adult) needs to possess to successfully recognise children: (1) ability to develop meaningful and in-depth relationship with children (i.e., familiarity with a particular child and their experience), (2) hermeneutic capacities (i.e., ability to make sense of indirect and non-verbal recognitional claims) and (3) ability to make prudential judgments (i.e., determining which aspects of the interaction are significant for the struggle for recognition). These discussions prepare ground for my second contribution—one pertaining to recognition and philosophy of technology. The interest in the application of Honneth's work to human-technology relations has been growing in recent years (Daus 2025; Waelen 2023; Waelen and Wieczorek 2022). However, the discussions so far have not considered children's experience of being (mis)recognised by technology and there are good reasons to suspect that algorithmic tools might not be able to engage in interpretation and make careful judgments necessary to adequately recognise children's perspectives. The increasing role that digital technologies play in children's lives—they now form a significant part of children's social relations, education and play—calls for the examination of their impact on children's self-development. The development of recognition theory and its application to children that I present in this chapter allows me to deepen our understanding of recognition by technology as applied to the youngest members of the society.

In what follows, I first briefly summarise Honneth's theory and its three modes of recognition—love, rights and solidarity (Sect. 2). I devote particular attention to love as Honneth presents his most extensive consideration of children's experiences of recognition while discussing primary relationships of care—which he considers paradigmatic examples of love. I combine this with insights from social work and child care literature to demonstrate how children experience (mis)recognition and to discuss the extent to which their struggle for recognition differs from adults (Sect. 3). This allows me to highlight characteristics of positive recognitional relationships between children and adults. In Sect. 4, I outline the limitations of contemporary algorithmic technologies in the context of recognition of children. I conclude by sketching an agenda for further research on recognition and children, particularly in the context of technology.

2 Honneth's Tripartite Framework of Recognition from a Child-Centric Perspective

In *The Struggle for Recognition* Honneth (1996) attempted to form an empirically grounded social theory that could explain the relational nature of our experience and account for the impact interactions with others have on the individual and social development. In his view, the experiences of being granted or denied recognition (and recognising others in turn) are fundamental for the formation of one's sense of self and fuel many social movements (e.g., as marginalised groups struggle to gain recognition from dominant forces).

Honneth's is not the only recognition theory as he himself has developed Hegel's early formulation of recognitional ideas and other thinkers have introduced their own theories of recognition (Butler 2008; Fraser 2000; Taylor 1994). However, Honneth's account is arguably the most influential and it owes much of its success to its Hegel-inspired tripartite division of relations of recognition—into love, rights and solidarity. According to this view, we are recognised and misrecognised by others on three distinct but complementary dimensions which correspond to our physical and emotional needs (love), our autonomy and capacity to participate in shared decision-making (rights), and our contributions to socially esteemed goals and practices (solidarity).

Love is arguably the most foundational of the three relations as it is a central feature of our first interactions with others and can be seen as informing all future experiences of recognition (Honneth 1996, p. 107; Smith et al. 2017). Honneth takes 'mother'-child relationships as the basis of all love relationships (Honneth 1996, 105), although he is careful to distinguish that he does not refer to 'mothers' in the literal sense, but uses inverted commas to denote a child's primary caregiver—a mother figure, so to speak.[1]

Drawing on Winnicot's psychoanalytical work on early childhood relationships (fitted within the overarching frame of Hegel's recognitional concepts and Mead's account of social development), Honneth observes that relationships between infant children and their 'mothers' are not uncomplicated. Rather, they develop within a "process of mutual recognition" (Honneth 1996, p. 98) in which the child and the 'mother' gradually learn to conceive of each other as independent beings with their own needs, desires and challenges. For the child, this experience demonstrates that they do not have boundless influence over the world (especially as during the first month of their life, a child does not fully delineate between themselves and external reality). Their needs and wants (hunger, comfort, contact, etc.) are not satisfied whenever they wish for that to happen but depend on a person external to them—a 'mother' over whom the child has no control and who can pursue their own wishes and desires. In healthy relationships, this is a moment when the child begins to develop a positive relationship to themselves.

If care and attention are not unconditional, the experience of having our needs recognised and acted upon teaches us that we are persons worthy of such care and attention, persons that matter in the eyes of the other—and Honneth labels this relation to self "basic self-confidence" (see Honneth 1996, p. 129, Fig. 2). However, experiences of being misrecognised within the sphere of love—e.g., through neglect or outright abuse and rape—are similarly constitutive and can threaten or even shatter one's sense of physical and emotional integrity. This is especially important as recognition is never a one-and-done process. To maintain a positive relation to oneself (and to others), we need constant care and attention—love—shared with others, throughout the course of our lives, not only during childhood. This explains how traumatic experiences (e.g., being a victim of violence) can destroy self-confidence built up over the course of a happy childhood (and later life). At the same time, this lifelong importance of love can help us understand why many social struggles are rooted in, account for and seek to alleviate experiences of deprivation and want (even if they are inseparable from the desire to assert one's rights and have one's social contribution adequately valued).

According to Honneth, relations of recognition are also characterised by mutuality. Primary relationships of care provide a good illustration of this

[1] Honneth (2014) also recognises that relations of love are not merely arrangements between two individuals, but are often institutionalised. The rights and obligations associated with marriage are a good example of this.

mutuality, as the 'mother's' sense of self is also transformed through the experience of nurturing a child. While the initial relationship is characterised by a sense of complete dependence—to the point that 'mothers' may experience a sense of failure due to the child's "helplessness and neediness" (Honneth 1996, p. 99)—its development requires the 'mother' to establish boundaries between them and the child. Following Winnicot, Honneth discusses that upon realisation of the independence of the 'mother', children often begin to bite, kick and otherwise struggle to test the distinction between them and this newly discovered external being. However, a similar process occurs for the 'mother', who needs to come to terms with the infant's own will and their existence as both a dependent object of care and an independent person in their own right. Love and care are fundamental human experiences (to the point where care ethicists categorise them as a primary way of being human, see Noddings 2002, 2003) and the relationship of love has a significant impact also on the 'mother's' self-development. To feel "the capacity for the shared experience of emotions and perceptions" (Honneth 1996, p. 97) can help the 'mother' conceptualise themselves as a person capable of providing love and care and to recognise the another person as someone worthy and in need of such love and care. It is also worth stressing that recognition in primary relationships of care occurs despite the vast communicative gap separating the 'mother' and the cared-for infant—just as the child struggles to identify their sense of want and to make it intelligible, the 'mother' wrestles with the difficulty of understanding the child's communicative efforts and responding to the needs that they express.

The discussion of love also makes it possible to highlight that relations of recognition are simultaneously mutual and asymmetrical (at least in the context of love, and as I argue further, in the experience of children). The child's experience of being recognised and the 'mother's' experience of recognising the child are distinct due to their psychological/developmental significance—the fuzzy feeling of having one's needs met is different from the fuzzy feeling of successfully providing care (and such asymmetry is present in all relationships of love). However, they also differ in the communicative challenges that they entail. The child faces the challenge of making their perspective intelligible, while successful recognition requires the 'mother' to possess not only a degree of good will but also relatively developed hermeneutic capacities. Such asymmetry exposes the one struggling for recognition to a great deal of dependence and vulnerability. Of course, such power asymmetries are present in all relationships, but, as I argue in the next section, children are particularly exposed to them.

Although Honneth devotes a great deal of attention to children's experience of recognition when discussing love—and he even considers children's early experiences of care and nurturing as constitutive for all other relationships of recognition—children largely disappear in his later account. In fact, Honneth discusses his move from love to rights and solidarity in parallel to the move from childhood to adolescence on the developmental plane. Discussing Mead's distinction between play (i.e., an imaginative activity in which a child makes up the patterns of conduct as they go along) and games (i.e., an activity with clearly specified rules that allow a child to determine and inhabit the actions of other players before they occur), Honneth equates attaining maturity (perhaps rightly so) with the attainment of the ability to relate one's conduct to the rules governing the conduct of a generalised other. In this way, he argues that "one becomes a socially accepted member of one's community" (Honneth 1996, p. 78) by guiding their behaviour in accordance with abstract norms—an ability that children distinctly lack, at least in this account.

The move from love to rights and solidarity is then precisely the point at which Honneth abandons a child's perspective and this makes him vulnerable to criticisms of adultism (Mitchell 2022; Thomas 2012; Warming 2015). I am not convinced that Honneth is at fault if the charge amounts to treating children only as adults but smaller—a mistake committed by many social philosophers. After all, he does delineate children's unique experiences of (mis)recognition while discussing love. However, it is true (and unfortunate) that children disappear in his discussions of rights and solidarity as he only considers these two relations from the perspective of fully-developed subjects. This is a glaring omission and a limitation of Honneth's recognition theory. Therefore I suggest that the theory needs to be adapted to fully account for the recognitional experience of children—and such an adaptation, in my view, paints a fuller picture of the relations of recognition shaping our society.

This limitation is particularly evident in Honneth's discussion of rights and solidarity as symmetrical, which stands in stark contrast to the relation of love, which, as discussed above, can be seen as necessarily asymmetric in Honneth's account, regardless of the identities of the parties in the relation. In principle, recognising the other on the plane of rights—as a person capable of making intersubjectively binding decisions, of following mutually agreed upon rules and contracts, and of demanding enforcement of such agreements—is indeed symmetrical. I (should) recognise you as a bearer of rights in the same way as I recognise myself and expect to be recognised in turn. However, even though children are bearers of rights, such as the right to self-determination (and other rights discussed, e.g., in the UN Convention on the Rights of the Child), we generally recognise that they are not bearers of rights in the same way adults are. For example, we put constraints on their decision-making capacities by not allowing them to, e.g., vote in elections or overruling their choices if we consider them dangerous or even inappropriate.

Similarly, for Honneth solidarity—the recognition of others' contribution to socially valued goals—is symmetrical, pure and simple (Honneth 1996, p. 129). There might be differences in the degree of esteem we attribute to each other—doctors, rightfully, are esteemed much more than investment bankers—but it is essentially the same *kind* of esteem that we grant and expect from one another. However, the solidarity we extend towards children is qualitatively different. We value their place in the society according to different rules than those applied to adults—for example, we might tend to look at a child's future potential rather than past and present accomplishments when extending praise. Perhaps for this reason authors applying Honneth's theory to child-centric empirical contexts notice that rights and solidarity are not as prevalent in children's experience as love—even though it is obvious that children experience esteem from parents, teachers and peers in the form of praise (Thomas 2012; Thomas et al. 2016; Warming 2006, 2015).

This is significant, because the differences in relationships of love between two adults and in relationships of love between 'mothers' and children seem to be differences in degree rather than differences in kind. Adults generally possess a greater degree of independence and self-reliance than children, but an adult in need (e.g., one that is hungry or homeless) is still in a vulnerable position relative to another adult. While the recognition there might be mutual (i.e., one adult is recognised as someone in need of care and recognises the other as a potential carer), it is not symmetrical (i.e., both parties need to recognise something completely different in each other).

Children's unique position in the struggles for recognition—one characterised by asymmetry and consequently vulnerability at all three planes of recognition—warrants better theorisation than the one presented by Honneth. Children still struggle to have their needs met, their voices heard and their accomplishments recognised. However, their struggle is qualitatively different than the one undertaken by adults. It entails different responsibilities and depends on different capacities on the part of those asked for recognition. For this reason, in the next section I provide a more detailed examination of children's experiences of (mis)recognition—drawing on empirical work done by social work and child care scholars inspired by Honneth's theory—and attempt to determine which attitudes and norms should guide those attempting to do justice to children's recognitional claims.

3 Towards a Child-Centric Account of Recognition

Have you ever asked a hungry child what they would like to eat only to dismiss their initial response and suggest a healthier or more nutritious alternative? Have you ever dismissed a child's tantrum or declarations of hate and sadness as them "just being tired"? Do you allow children in your care to make decisions for themselves but overrule their choices if you consider them harmful or irresponsible? Have you ever praised a hideous portrait of yourself only because it was drawn by a six-year-old?

These and many similar examples illustrate that the recognition of children does not follow the same logic as the recognition of adults. Yes, we have a responsibility to account for children's expressed needs and desires, to respect their choices and to adequately value their accomplishments. However, this must be balanced with the fact that children often do not know what is good for them—or that adults know it better.

Admittedly, since recognitional claims are part of a *struggle* for recognition, there are always disagreements on how to interpret expressions of those participating in the struggle or how much recognition is due to different subjects. In many instances, children may face challenges with expressing their claims accurately or they may not be able to fully interrogate their needs and wishes—after all, the process of growing up involves learning who they are and what they want.[2] For this reason, if we want to foster children's self-confidence, self-respect and self-esteem, we might at times go against what they literally express in their recognitional claims.

As children are minors—not fully developed subjects—they typically enter recognitional relations from a position of vulnerability and see their agency denied for reasons both well-meaning and malicious. These may include the desire to protect them from negative consequences of their choices or the desire to foster their actual rather than expressed needs. However, such interventions on the part of adults can also be a result of paternalism, the exercise of power and control, or misguided ideas about children's role and freedom. In this context, Warming's (Warming 2015, p. 251) cautions against instances in which "the overruling and patronizing attitude of adults appears as an expression of care and adult responsibility". Elsewhere, while discussing the experience of foster children with political participation, she reports on feelings that "they only have a say as long as their perceptions and desires are in

[2] Admittedly, communication and introspection can be challenging for adults as well.

accordance with – or at least not significantly at odds with those that the professionals identify as the best interests of the child" (Warming 2006, p. 42).

Foster children's dependence on adults who are often not in close emotional relationships with them (e.g., case workers who might be responsible for many children at the same time) makes them especially vulnerable, but it is a vulnerability experienced by children all over the world. It is often the responsible adults who are tasked with determining what is the best interest of the child and Mitchell (2022, p. 527) discusses such determinations as discretionary—hinting at the degree of power adults wield over the children under their care—but I argue that it would be appropriate to discuss such decisions as *prudential.* I mean by this that the recognition of children[3] involves a great deal of practical wisdom and hermeneutic capacity—giving appropriate regard and making sense of myriad contextual, emotional, half-expressed and unacknowledged cues that provide a fuller picture of recognitional needs than their direct expressions.

Interestingly, Honneth is open to acknowledging that struggles for recognition do not always play out through well-articulated and comprehensive recognitional claims. He argues that expressions of outrage might be good indicators of situations where due recognition is missing as "the negative emotional reactions accompanying the experience of disrespect could represent precisely the affective motivational basis in which the struggled-for recognition is anchored" (Honneth 1996, p. 135). In an earlier essay, *Moral Consciousness and Class Domination: Some Problems in the Analysis of Hidden Morality* (Honneth 2007), he also gives an in-depth assessment of the unacknowledged moral convictions of the dominated classes. In his view, those who do not possess the communicative and hermeneutic capacities or the means of communication (e.g., presence in widely circulated media outlets) might be deprived of opportunities to formulate a clear expression of their moral positions—in fact, since they are never asked, they might never do so. However, Honneth argues that a lack of articulate and consistent views does not mean that dominated classes do not possess their morality—they do not merely parrot the moral sentiments expressed by others. In fact, for Honneth it is their expressions of indignation that are more informative of their moral positions than any positive formulations of ethical judgments. For example, he traces how members of the working class develop a "counterculture of compensatory respect"—a set of beliefs and practices that serve to acknowledge the value of work that does not receive appropriate esteem from the other strata of the society.

---

[3] And to all lesser degree, all kinds of recognition.

I argue that in the context of children's struggles for recognition we should also attempt to look at recognitional claims that do not find a clear and coherent linguistic expression. To put it simply, recognition of children needs to acknowledge that children think and feel more than they can name and express. Precisely, recognitional work on the part of adults might involve identifying half-formulated claims and helping children find the appropriate means of formulating them.

A good example of this is offered by Thomas et al. (2016), who studied the affective and communicative efforts undertaken by teachers to promote their students' wellbeing and framed their findings within Honneth's theory of recognition. They argue that teachers respond to children's emotional and physical needs by first developing a close relationship and establishing trust. This allows them to notice signs of distress and act proactively even in the absence of clear verbal communication—for example, to spot that a student has not been themselves due to a death in the family and offer appropriate support. In a sense, children first need to feel understood before adults can attempt to make sense of their claims for recognition.

Such patterns of close relations of friendship, partnership and care are prevalent in recognition-inspired empirical literature, with authors who discuss children's political participation (Thomas 2012; Warming 2006) and children's engagement with care workers (Mitchell 2022; Paulsen and Thomas 2018; Warming 2015) highlighting the personal engagement and familiarity as fundamental to successful engagement with children. As they point out, adults who maintain close relationships with children under their care are not only able to pick up on emotional and context cues but also elicit more open, honest and direct communication from their wards—especially in the context of children who are already deprived of uplifting interactions with supportive adults (Warming 2006).

Moreover, as pointed out by Häkli et al. (2018, p. 7), getting to know children allows care workers to offer their support in a much more nuanced way, for example by recognising what "they themselves consider significant in their current lives". This allows them not only to engage with their wards in a much more meaningful way—i.e., without merely pasting their own sense of worth over the experience of the child—but also demonstrate that they understand the child's point of view without necessarily endorsing or condemning the acts or moral judgments discussed by the child. This is significant as it allows them to raise controversial issues without attributing blame that might threaten the relationship.

Similar matters are discussed by Houston and Dolan (2008) and Smith et al. (2017) in their Honneth-inspired frameworks for incorporating recognitional concepts in care workers' and child care professionals' arsenal of tools. Their work suggests that the maintenance of close relationships with children under care—ones based on in-depth knowledge of a child's interests, challenges and personal circumstances—allows responsible adults to offer more contextualised and concrete support than mere acknowledgment of, e.g., needs and wants. Consequently, according to Houston and Dolan (2008) the concept of recognition helps highlight that knowing another person makes it possible to offer support and behave in a way specifically aligned to them.

Inspired by such discussions, I propose three factors fundamental to successful recognition of children. First, the recognising party needs to be capable of establishing a relationship with the recognised child that is based on familiarity (i.e., knowledge of the child), friendship (or other positive emotional bonds), and trust. For the purposes of this discussion, I suggest labelling this trait as sociability. Second, the recognising party needs to be able to interpret cues that are non-verbal or even not explicitly communicated over the course of the interaction—these may include emotional reactions, changes to customary behaviour, or specificities of a child's personal circumstances. This ability could be described under the umbrella term of hermeneutic capacity. Third, the recognising party needs to be capable of making judgments about the significance of the factors just discussed. For example, an experienced care worker or teacher might know from experience *which* contexts are likely to factor into a child's need for recognition, and which can be dismissed as background noise. The ability to make such judgments could be understood as a form of practical wisdom.

Of course, people possessing a degree of sociability, hermeneutic capacities and practical wisdom are likely to be much more successful in granting appropriate recognition in all manners of interactions, not just those involving children. However, due to the complex nature of children's recognitional needs and claims—characterised by a greater degree of asymmetry and vulnerability—I merely seek to argue that these capacities are much more important in child-centric contexts.

4 Children, Recognition and Technology

While care workers, teachers or even parents generally possess or are able to acquire and develop such capacities, today children are increasingly using algorithmic tools in education, play and everyday social interactions. These tools, as I argue in this section, do not possess the proposed capacities and are unlikely to successfully simulate them. Consequently, while recognition-related literature has already identified some deficiencies in technologically-mediated recognition (Daus 2025; Waelen 2023; Waelen and Wieczorek 2022), my analysis allows me to demonstrate that such issues are likely to be exaggerated in the context of the (mis)recognition of children by technology.

If sociability is conducive to the successful recognition of children, the possibility of being recognised by a technology is likely to depend on technology's ability to establish in-depth relationships with humans. Consequently, it is appropriate to ask which aspects of sociability can be emulated by technology—namely, the ability to acquire in-depth knowledge of the interaction partner, the ability to forge positive emotional bonds and the ability to establish trust. I am not convinced that contemporary algorithmic technologies—such as personalisation algorithms or AI agents that have previously been discussed by recognition scholars—can succeed in that regard.

First, it is unclear whether the technologies in question can know anything. It has been suggested that AI tools, like large language models, are indifferent towards truth (Hicks et al. 2024), that statements produced by them do not demonstrate understanding or refer to real-world objects in the way that human sentences do (Coeckelbergh and Gunkel 2023), and that they merely approximate plausibly sounding text through statistical modelling (Bender et al. 2021).

Similarly, although AI and other algorithmic tools process personal information, having access to someone's digital data is not necessarily the same as getting to know them. Data collected by algorithms is typically fragmented (it does not contain the entirety of relevant information), often inaccurate (due to the imprecision of the means of data collection), and always reductive (since complex phenomena, such as gender identity, might be represented in reductive categories). For these reason, researchers have argued that it might be most appropriate to talk not about personal data, but *data doubles*—digital representations whose characteristics largely overlap but are not identical with the characteristics of the represented person (Haggerty and Ericson 2000; Ruckenstein 2014; Wieczorek 2024). And as I discuss elsewhere (Wieczorek 2024), the mediation of interpersonal interactions through data introduces a degree of distance which complicates the possibility of getting to know and becoming close with the datafied person. Consequently, even if we accept that AI agents or personalisation algorithms may *know* us or something about us, they would do so in qualitatively different ways than humans—and I am sceptical whether such knowledge would amount to the knowledge caregivers have about their wards.

Moreover, in recent years, human-robot relationships have been a subject of close scrutiny and it is worth noting that even authors discussing hypothetical, highly advanced social robots are unable to determine whether such relationships are feasible (Gunkel 2018; Nyholm and Frank 2017; Nyholm and Smids 2020) and the consensus seems to be that human-robot relationships would, at the very least, be qualitatively different than human ones (Ryland 2021). However, those examining technologies already deployed in the society adopt more sceptical views (Elder 2017; Fröding and Peterson 2021). Moreover, the recognition-conducive relationships I highlighted in the previous section are characterised by emotional investment on the part of the adult carers—one that is recognised by the cared-for child. Even if human-machine relationships can approximate some aspects of such positive bonds, I believe that they would falter in recognitional terms due to the machine's incapacity to feel emotions (and since relations of recognition are mutual, the child's perception of this incapacity).

As Fröding and Peterson (2021) argue, AI algorithms are incapable of genuine friendship because they lack the ability to reciprocate. This is not merely a technological limitation as even if AI tools are developed to mimic a relation virtually indistinguishable from a human friendship—what Fröding and Peterson call as-if-friendship—users' awareness of their interaction with an AI would elicit an altogether different response. For example, Andries and Robertson (2023) demonstrate that even when children anthropomorphise AI assistants and overestimate their intelligence and agency, they are aware that such AI agents possess a different status than humans and this awareness influences children's behaviour (in this case, children were treating the assistant with additional politeness—one distinct from politeness offered to humans).[4]

This implies that the awareness of the artificial nature of AI agents may prompt children to treat them with a degree of distance, raising questions about whether they would reciprocate any simulated emotional bonds or whether they would trust artificial agents in the same way they trust human caregivers. Consequently, while algorithmic technologies might be able to approximate some form of sociability, it is not clear that it would be the kind of sociability that would be helpful in recognising children—e.g., prompt them to share their experiences and perspectives in a way that enables greater understanding of their recognitional claims.

---

[4] See also Sam Altman's plea for users not to thank ChatGPT to limit computing costs: https://www.nytimes.com/2025/04/24/technology/chatgpt-alexa-please-thank-you.html.

I am sceptical that such simulated sociability would succeed in prompting children to disclose their recognitional needs, allow them to become vulnerable in a manner that is comfortable to them, and help them communicate their experiences in a more accessible and coherent way (as happens when human carers develop relationships with children). However, examples of AI companions show that some people seek to maintain relationships with algorithmic tools and are open to share deeply personal and emotional details—for the purpose of this argument, let us assume that such disclosures are functionally similar to those prompted by human caregivers.

As noted above, successful recognition of children's claims depends not only on their ability, comfort and willingness to communicate their needs and wants, but also on the hermeneutic capacities of the recognising party. I already highlighted that it is doubtful whether algorithmic tools can know anything, but it is also doubtful whether they can process and classify information that is relevant to recognitional claims. For example, machines' ability to recognise emotion has been a topic of intensive discussion among philosophers and Waelen (2024) recently demonstrated that such functionalities are typically based on reductive understandings of emotions and often reproduce biased judgments of those in power. Similarly, algorithmic tools have previously been demonstrated to misrecognise non-verbal communicative cues such as facial expressions and tone of voice—as Waelen and I discussed before (Waelen and Wieczorek 2022).

The translation of complex phenomena such as emotions, intentions, wants and desires into quantified and machine-readable data is inherently reductive. Embodied, situated and sometimes poorly understood experiences get their edges shaved off to fit narrow categories and proxies. And while such data points can be compared between users and aggregated to offer statistical significance and basis for prediction, they obfuscate the specificity of individual circumstances and remove many contextual details that are fundamental for meaningful recognition. If recognition involves interpretation—a process of responding to and making sense of often minute and less than straightforward details—algorithmic tools often engage in the opposite. The aggregation of ruthless pursuit of statistically significant patterns in the supplied data might be better framed as a brute force approach to understanding.

There abound examples that children's needs, perspectives and emotions are particularly overlooked by algorithms, as discussed for example in the literature on educational AI which points out how such tools overlook children's emotional expressions and the complexities of their speech (Leaton Gray 2020; Schiff 2021; Selwyn 2022), while also enforcing rigid standards of behaviour and failing to recognise non-verbal expressions of those who fail to conform to such standards (Madaio et al. 2022). Consequently, I argue that algorithmic tools are likely to struggle precisely in the hermeneutic work I discussed in the previous section by highlighting the interpretative efforts of human caregivers. More precisely, such tools might not be able to engage in hermeneutic work at all—especially if we consider the capacity to undertake such work as uniquely human, as discussed by hermeneutic philosophers.

Consequently, as algorithmic tools may be unable to "understand" the variety of cues through which children might express or signal their recognitional claims (e.g., emotions, body language, tone of voice, ambiguous and misleading verbal communication), they will certainly fail at determining which of these cues are significant. Naturally, such tools have demonstrated great prowess at identifying statistical patterns which could be understood to demonstrate when children feel distressed, neglected or otherwise unhappy. However, the ability to flag that some behaviour occurs is quite distinct from the ability to flag *why* it does. For example, the literature on educational AI highlights that tools that are successful at flagging problem behaviour do not account for contextual information that might help educators understand the reasons for such behaviour (Gillani et al. 2023; Rowe 2019).

Significantly, teachers often engage in a variety of prudential judgments when evaluating student behaviour. In interviews with Kousa and Niemi (2023), teachers point out that what happens in class is often dependent on a wide range of factors that cannot be accounted for by algorithms (e.g., teachers know that students are more disruptive in bad weather). Similarly, in their analysis of the practices of an Australian startup that sought to automate attendance taking through computer vision technology, Selwyn et al. (2023) point out that even during seemingly mundane practices such as roll call teachers routinely exercise judgments that help them understand and relate to their students (e.g., by noticing signs of distress, confusion, or inattentiveness). Such prudential judgments are also central to assessment practices which offer educators opportunities to get to know their students, learn the particularities of their writing voice, identify interests, or understand struggles and problems. In this context, much of a teacher's activity involves determining which aspects of a student's work are significant and which could be dismissed as background noise.

As I argue, such judgments should be understood as prudential—e.g., exercises of practical wisdom. Admittedly, some philosophers have extensively discussed machines' ability to simulate a form of practical wisdom and prescribed design interventions that might help developers achieve such goals (Sullins 2025). However, I understand the practical wisdom exemplified by teachers and care workers as inherently linked to their experience—both their experience of working with and recognising children and their experience in recognitional struggles and the expression of recognitional claims.

Consequently, the ability to make the appropriate judgments in the context of recognition of children cannot be acquired by crunching more and more data about them—or, similarly, by only reading more and more books about their experiences and development. If we take seriously Honneth's claims about the foundational nature of 'mother'-child relationships for all relations of recognition, it might be prudent to conclude that the ability to recognise children rests on two pillars—that of being with children and that of being (or once having been) a child. This insight is perhaps best expressed by the novelist Ian McEwan in his *Machines Like Me*. There, a humanoid robot Adam is quite proficient in interacting with humans—and even engaging in a love triangle with two of them—but cannot relate to children. In a memorable scene, he attempts to, and fails, to engage in child's play (here, note my earlier discussion of Honneth's and Mead's distinction between play and games and AI's dependence on the recognition of reliable and predictable patterns that are absent from play). Adam was never a child, and he is only able to operate within a world of clearly established and easily graspable rules and boundaries. This world could not be more different than the world of play and, consequently, the world of children. Adam's faults in this regard lie less in his never having been a child (who could blame the robot for this?) and more in his inability to be child*like*.

5 Conclusion

This chapter offered a critical examination of Honneth's recognition theory from the perspective of children's experiences of (mis)recognition. According to Honneth, early care relationships between infant children and their primary caregivers are foundational for all later relations of recognition. As I argued, the experience of having one's physical and emotional needs met despite the communicative gaps existing between the child and the 'mother' finds parallels in other dimensions of recognition involving children. In my account, Honneth recognises the asymmetry and vulnerability inherent to children's experience of love—children (and adults that are being cared for) are not only dependent on their carers but also receive a different kind of recognition than the recognition given to the carer. However, Honneth considers recognition relations at the level of rights and solidarity to be symmetrical, meaning that both recognising parties are (at least in ideal terms) due the same kind and level of respect and the esteem they exchange may differ in quantity but is qualitatively the same. This is not the experience of children who are vulnerable as bearers of rights and members of communities because their experiences of (mis)recognition are asymmetrical also in these contexts. As I note in Sect. 3, their ability to make binding decisions is constrained by responsible adults, and the valuation of their accomplishments follows a different logic. Consequently, due to the asymmetry and vulnerability inherent to children's experiences of (mis)recognition, I argued that successful and appropriate regard for their needs, legal personhood and societal contributions requires the recognising parties to bridge communicative gaps and exercise prudential judgments as part of the balancing act between restrictive paternalism and undifferentiated treatment of children as merely miniature adults.

Building on social work and childcare literature, I argued that recognition of children needs to acknowledge that children think and feel more than they can name and express. Children may often struggle to identify and name their experiences, which limits their ability to successfully engage in recognitional struggles and often calls for support of responsible adults. I highlighted that experienced care workers engage in relationship-building efforts to get to know their wards and thus better understand both the specificity of their needs and circumstances, as well as the means through which these might be discussed—which often includes learning to identify and interpret a child's non-verbal and indirect cues such as emotional reactions, references made in passing, etc.

On this basis, I argued that successful and appropriate recognition of children depends on the recognising party's ability to forge meaningful and deep personal relationships (sociability), their ability to interpret non-verbal and indirect forms of communication (hermeneutic capacity) and the ability to determine which parts of the communicative relationship are pertinent to a given struggle for recognition (practical wisdom). Over the course of Sect. 4, I demonstrated that algorithmic technologies demonstrate significant deficiencies in each of these aspects of recognition. Firstly, they are unable to form deep relationships—and if they do form some kind of relationship, a child's awareness of interacting with AI might influence what is disclosed and how. Secondly, they struggle with identifying emotions and accounting for context. Thirdly, they do not possess the experience necessary to make prudential judgments. For these reasons, I conclude that today's algorithmic tools are poorly equipped for recognising children at all three levels inherent to Honneth's theory—love, rights and solidarity.

However, as the growing discussion of technology and recognition demonstrates, the recognitional deficiencies of algorithms do not stop people from deploying such limited tools in contexts where experiences of (mis)recognition (by technology or not) might have significant constitutive impacts—i.e., shape our sense of self. For this reason, I suggest that future research on technology and recognition should examine specifically the situations in which children are the most likely to engage with algorithmic technologies and experience (mis)recognition. I already referred to some examples from critical "edtech" (educational technology) studies and would encourage recognition scholars to pay greater attention to the intersection of education and technology. However, other areas that warrant attention include children's experiences of the algorithms embedded in social media platforms (a major source of anxiety and self-doubt today), of the increased digitalisation of the provision of healthcare and social services, or of virtual spaces and virtual personas (e.g., through video games and interactions with AI companions). The efforts of social work and child care scholars, which were instrumental for the writing of this chapter, demonstrate that recognition theory has much to learn from and much to offer for the study of children and the complex nature of their struggles for recognition. Consequently, I believe that the inclusion of children's perspectives (perhaps also through empirical research) would be fundamental for the further development of our understanding of the relationship between recognition and technology.

References

Andries, Valentina, and Judy Robertson. 2023. Alexa Doesn't have that many feelings: Children's understanding of AI through interactions with smart speakers in their homes. *Computers and Education: Artificial Intelligence* 5:100176. https://doi.org/10.1016/j.caeai.2023.100176.

Bender, Emily M., Timnit Gebru, Angelina McMillan-Major, and Shmargaret Shmitchell. 2021. On the dangers of stochastic parrots: Can language models be too big? In *Proceedings of the 2021 ACM conference on fairness, accountability, and transparency*, 610–623. Virtual Event Canada: ACM. https://doi.org/10.1145/3442188.3445922.

Butler, Judith. 2008. Taking Another's view: Ambivalent implications. In *Reification: A new look at an old idea*, edited by Axel Honneth and Martin Jay, 97–119. The Berkeley Tanner Lectures. Oxford: Oxford University Press.

Coeckelbergh, Mark, and David J. Gunkel. 2023. ChatGPT: Deconstructing the debate and moving it forward. *AI & SOCIETY*. https://doi.org/10.1007/s00146-023-01710-4.

Daus, Zachary. 2025. Distribution, recognition, and just medical AI. *Philosophy & Technology* 38 (1): 11. https://doi.org/10.1007/s13347-025-00842-3.

Elder, Alexis. 2017. Robot friends for autistic children: Monopoly money or counterfeit currency? In *Robot ethics 2.0: From autonomous cars to artificial intelligence*, edited by Patrick Lin, Keith Abney, and Ryan Jenkins. Oxford University Press. https://doi.org/10.1093/oso/9780190652951.003.0008.

Fraser, Nancy. 2000. Rethinking recognition. *New Left Review* 3 (June): 107–120.

Fröding, Barbro, and Martin Peterson. 2021. Friendly AI. *Ethics and Information Technology* 23 (3): 207–214. https://doi.org/10.1007/s10676-020-09556-w.

Gillani, Nabeel, Rebecca Eynon, Catherine Chiabaut, and Kelsey Finkel. 2023. Unpacking the "black box" of AI in education. *Journal of Educational Technology & Society* 26 (1): 99–111.

Gunkel, David J. 2018. *Robot rights*. The MIT Press. https://doi.org/10.7551/mitpress/11444.001.0001.

Haggerty, Kevin D., and Richard V. Ericson. 2000. The Surveillant assemblage. *The British Journal of Sociology* 51 (4): 605–622. https://doi.org/10.1080/00071310020015280.

Häkli, Jouni, Riikka Korkiamäki, and Kirsi Pauliina Kallio. 2018. "Positive recognition" as a preventive approach in child and youth welfare services. *International Journal of Social Pedagogy* 7 (1): https://doi.org/10.14324/111.444.ijsp.2018.v7.1.005.

Hicks, Michael Townsen, James Humphries, and Joe Slater. 2024. ChatGPT is bullshit. *Ethics and Information Technology* 26 (2): 38. https://doi.org/10.1007/s10676-024-09775-5.

Honneth, Axel. 1996. *The struggle for recognition: The moral grammar of social conflicts*. 1st MIT Press ed. Studies in Contemporary German Social Thought. Cambridge, Mass: MIT Press.

Honneth, Axel. 2007. Moral Conciousness and class domination. In *Disrespect*. Cambridge: Polity Press.

Honneth, Axel. 2012. *The I in we: Studies in the theory of recognition*. Cambridge: Polity press.

Honneth, Axel. 2014. *Freedom's Right: The Social Foundations of Democratic Life*. Trans. G. Joseph. Cambridge: Polity.

Houston, Stan, and Pat Dolan. 2008. Conceptualising child and family support: The contribution of Honneth's critical theory of recognition. *Children & Society* 22 (6): 458–469. https://doi.org/10.1111/j.1099-0860.2007.00126.x.

Kousa, Päivi, and Hannele Niemi. 2023. Artificial intelligence ethics from the perspective of educational technology companies and schools. In *AI in learning: Designing the future*, edited by Hannele Niemi, Roy D. Pea, and Lu Yu, 283–296. Cham: Springer International Publishing. https://doi.org/10.1007/978-3-031-09687-7.

Leaton Gray, Sandra. 2020. Artificial intelligence in schools: Towards a democratic future. *London Review of Education*. https://doi.org/10.14324/LRE.18.2.02.

Madaio, Michael, Su Lin Blodgett, Elijah Mayfield, and Ezekiel Dixon-Román. 2022. Beyond "fairness": Structural (in)justice lenses on AI for education. In *The ethics of artificial intelligence in education*, edited by Wayne Holmes and Kaśka Porayska-Pomsta. New York: Routledge.

Mead, George Herbert. 2015. In *Mind, self, and society*, edited by Charles W. Morris, Daniel R. Huebner, and Hans Joas. Chicago; London: University of Chicago Press.The definitive edition

Mitchell, Mary. 2022. "Because I'm a kid ...": The struggle for recognition of children and young people involved in child and family social work. *Child & Family Social Work* 27 (3): 526–534. https://doi.org/10.1111/cfs.12905.

Noddings, Nel. 2002. *Educating moral people: A caring alternative to character education*. New York London: Teachers College Press.

Noddings, Nel. 2003. *Caring: A feminine approach to Ethics & Moral Education*. 2nd ed. Berkeley: University of California Press.

Nyholm, Sven, and Lily Eva Frank. 2017, October. *From Sex Robots to Love Robots: Is Mutual Love with a Robot Possible?*. https://doi.org/10.7551/mitpress/10718.003.0019.

Nyholm, Sven, and Jilles Smids. 2020. Can a robot be a good colleague? *Science and Engineering Ethics* 26 (4): 2169–2188. https://doi.org/10.1007/s11948-019-00172-6.

Paulsen, Veronika, and Nigel Thomas. 2018. The transition to adulthood from care as a struggle for recognition. *Child & Family Social Work* 23 (2): 163–170. https://doi.org/10.1111/cfs.12395.

Rowe, Michael. 2019. Shaping our algorithms before they shape us. In *Artificial intelligence and inclusive education: Speculative futures and emerging practices*, edited by Jeremy Knox, Yuchen Wang, and Michael Gallagher, 151–163. Perspectives on Rethinking and Reforming Education. Singapore: Springer. https://doi.org/10.1007/978-981-13-8161-4_9.

Ruckenstein, Minna. 2014. Visualized and interacted life: Personal analytics and engagements with data doubles. *Societies* 4 (1): 68–84. https://doi.org/10.3390/soc4010068.

Ryland, Helen. 2021. It's friendship, Jim, but not as we know it: A degrees-of-friendship view of human–robot friendships. *Minds and Machines* 31 (3): 377–393. https://doi.org/10.1007/s11023-021-09560-z.

Schiff, Daniel. 2021. Out of the laboratory and into the classroom: The future of artificial intelligence in education. *AI and Society*. https://doi.org/10.1007/s00146-020-01033-8.

Selwyn, Neil. 2022. The future of AI and education: Some cautionary notes. *European Journal of Education*. https://doi.org/10.1111/ejed.12532.

Selwyn, Neil, Liz Campbell, and Mark Andrejevic. 2023. Autoroll: Scripting the emergence of classroom facial recognition technology. *Learning, Media and Technology* 48 (1): 166–179. https://doi.org/10.1080/17439884.2022.2039938.

Smith, Mark, Claire Cameron, and Daniela Reimer. 2017. From attachment to recognition for children in care. *The British Journal of Social Work* 47 (6): 1606–1623. https://doi.org/10.1093/bjsw/bcx096.

Sullins, John P. 2025. Automated ethical practical reasoning: The problem of artificial Phronesis. In *Robophilosophy—Philosophy of, for, and by social robotics*, edited by Johanna Seibt, Raul Hakli, and Marco Norskov. MIT Press.

Taylor, Charles. 1994. The politics of recognition. In *Multiculturalism: Examining the politics of recognition*, edited by Nachdr Amy Gutmann. Princeton, NJ: Princeton Univ. Press.Princeton Paperbacks

Thomas, Nigel. 2012. Love, rights and solidarity: Studying Children's participation using Honneth's theory of recognition. *Childhood* 19 (4): 453–466. https://doi.org/10.1177/0907568211434604.

Thomas, Nigel, Anne Graham, Mary Ann Powell, and Robyn Fitzgerald. 2016. Conceptualisations of Children's wellbeing at school: The contribution of recognition theory. *Childhood* 23 (4): 506–520. https://doi.org/10.1177/0907568215622802.

Waelen, Rosalie. 2023. The struggle for recognition in the age of facial recognition technology. *AI and Ethics* 3 (1): 215–222. https://doi.org/10.1007/s43681-022-00146-8.

Waelen, Rosalie. 2024. Philosophical lessons for emotion recognition technology. *Minds and Machines* 34 (1): 3. https://doi.org/10.1007/s11023-024-09671-3.

Waelen, Rosalie, and Michał Wieczorek. 2022. The struggle for AI's recognition: Understanding the normative implications of gender bias in AI with Honneth's theory of recognition. *Philosophy & Technology* 35 (2): 53. https://doi.org/10.1007/s13347-022-00548-w.

Warming, Hanne. 2006. "How can you know? You're not a Foster child": Dilemmas and possibilities of giving voice to children in Foster Care. *Children, Youth and Environments* 16 (2): 28–50. https://doi.org/10.1353/cye.2006.0016.

Warming, Hanne. 2015. The life of children in Care in Denmark: A struggle over recognition. *Childhood* 22 (2): 248–262. https://doi.org/10.1177/0907568214522838.

Wieczorek, Michał. 2024. Community, solidarity and care through data? An ethical analysis of the interpersonal dimension of self-tracking. *AI & SOCIETY*. https://doi.org/10.1007/s00146-024-02013-y.

Winnicott, Donald W. 2018. *The maturational processes and the facilitating environment: Studies in the theory of emotional development*. London: Routledge. https://doi.org/10.4324/9780429482410

R-E-S-P-E-C-T: Find Out What It Means to AI Policy

Caroline Stitt

Abstract This chapter critiques the European Union's *Artificial Intelligence Act* (AI Act) through the lens of normative recognition in the form of respect, as received from institutions through policy and law. Drawing on Honneth's understanding of respect as a system of individual rights, Charles Taylor's notion of the 'politics of universalism' and Nancy Fraser's principle of 'parity of participation', the chapter explores to what extent the Act enables individuals' autonomy and self-development by recognizing them as bearers of rights who ought to be treated equally to other members of society. It is shown that the AI Act demonstrates an intent to respect individuals by protecting fundamental rights and promoting human autonomy through, for example, human oversight and transparency provisions and prohibitions on manipulative AI practices. However, as is argued in this chapter, this intent of the AI Act to respect individuals falls short in practice due to the Act's predominantly risk-based approach coupled with weak enforcement mechanisms, industry self-regulation, and significant loopholes—particularly in law enforcement, migration, and national security contexts—which undermine its ability to enable individuals to enjoy full rational autonomy and develop self-respect. Moreover, the Act largely fails to protect the rights of individuals located outside the EU as well as vulnerable populations such as migrants and asylum

C. Stitt (✉)
AI + Planetary Justice Alliance, Berlin, Germany
e-mail: caroline@aiplanetaryjustice.com

R. Waelen (ed.), *Normative Relations of Recognition in the Age of AI*, Philosophy and Ethics of AI 2, https://doi.org/10.1007/978-3-032-34028-3_8

seekers, whom the Act does not recognize as equal bearers of rights. Ultimately, despite its laudable intentions, the AI Act does not meet the normative condition of respect as enabling autonomy through law.

Keywords EU AI act • Recognition theory • AI policy • Respect • Rights-based approach • Risk-based approach

1 Introduction

According to G.W.F. Hegel and Axel Honneth's tripartite definition, respect is a form of recognition that is received by persons from societal institutions. What is recognized through respect is that individuals are bearers of rights, who ought to be treated equally to other members of society. The underlying reasoning is that all persons are owed respect by virtue of the moral status that is intrinsic to their humanity and personhood. In other words, respect is owed to all persons, regardless of any individual traits. While the moral significance of universal respect is discussed in many political-philosophical traditions (among others by John Rawls in his famous *A Theory of Justice* (1971)), recognition theories are unique in that they take the moral relevance of recognition as respect to exist in the fact that receiving respect is a requirement for a positive relation-to-self. More precisely, it is through receiving institutional respect for one's moral rights and equal worth that a person is able to develop self-respect. Hence, according to recognition theory, respecting subjects through policy and law is about more than the liberal moral principle of fair and equal treatment or the intrinsic value of autonomy—it is about the impact of societal institutions on individuals' psychological development and wellbeing. Institutions ought to respect human autonomy through policy and law because doing so actually enables autonomy.

The aim of this chapter is to explore to what extent AI policy can be said to respect individuals through enabling their autonomy and thus their positive relation-to-self. The chapter focuses specifically on the European Union's Artificial Intelligence Act (hereafter 'AI Act'), which came into force in August 2024, and which offers the first comprehensive legal framework governing artificial intelligence. Fundamental rights (such as human dignity, non-discrimination, and respect for private life) are the foundation of the EU's guiding normative framework for regulating AI—an approach recommended by the High-Level Expert Group on Artificial Intelligence, which the Commission convened to provide advice on its AI strategy (Smuha et al. 2021). Fundamental rights were chosen by the Group because of their universality–they are recognized by nearly every country in the world, and are universally applicable to all individuals. As such, the Group argued that "respect for fundamental rights, within a framework of democracy and the rule of law, provides the most promising foundations for identifying abstract ethical principles and values, which can be operationalized in the context of AI" (AI HLEG 2019). To a certain extent, this intent to respect was enshrined in the final text. Article 1 of the Act, for example, clearly sets out its mandate to ensure "a high level of protection of health, safety, fundamental rights [...] against the harmful effects of AI systems in the Union" (European Union, 2024).

This chapter will investigate to what extent the AI Act lives up to its laudable intention to respect individuals, by critically analyzing whether the AI Acts effectively enables human autonomy in practice. It will do so through a three-part analysis. The first section will explore the ways in which the Act intends to respect individuals, through prohibitions on harmful AI practices as well as human oversight and transparency provisions. The second section will critically analyze to what extent the Act succeeds in realizing its institutional goals of protecting fundamental rights and to what extent it can be said to recognize individuals in practice through enabling human autonomy. The last section will offer some normative considerations of the role of recognition and respect in AI policy.

In its understanding of recognition as respect, this chapter draws on the general understanding of respect as "a universal form of recognition due to humans, for features that they have in common with all of their fellows" (Thompson 2006, p. 15). While recognition scholars like Charles Taylor, Nancy Fraser, and Axel Honneth disagree on a range of issues, including the grounds for respect and the extent to which respect should be tied to rights, they have this general understanding of respect in common. The basic principle of respect as equal rights underpins Taylor's notion of the politics of universalism, Fraser's principle of parity of participation, and Honneth's understanding of respect as a system of individual rights (Thompson 2006). In other words, the shared perspective is that all individuals must enjoy the same fundamental rights in order to exercise their human autonomy.

2 An Intent to Respect

This section explores to ways in which the AI Act demonstrates, both implicitly and explicitly, an *intent to respect* individuals through protecting their fundamental rights and enabling their rational autonomy. The provisions explored demonstrate an intent to respect individuals' capacity for rational autonomy in terms of its Kantian interpretation, which underpins both Taylor and Honneth's understandings of respect. Taylor's politics of universalism posits that respect is due to individuals in light of their 'universal human potential' for autonomy, or what Taylor refers to as 'our status as rational agents'. In other words, as Simon Thompson argues, "It is because people are able to exercise rational control over their lives that they should be treated with respect" (Thompson 2006, p. 45). Similarly, Honneth regards 'moral responsibility as the respect-worthy core of the person' and contends that it is because humans have a capacity for rational autonomy that they can have moral responsibility attributed to them. Hence to respect others is to regard them as 'capable of acting autonomously on the basis of rational insight' (Thompson 2006, p. 48).

2.1 Risk-Based Approach

The AI Act takes a risk-based approach to regulating AI. This means that it categorizes AI systems in relation to their level of risk to the fundamental rights and establishes specific requirements for each risk category that deployers, providers and other stakeholders must comply with. Systems deemed an 'unacceptable' risk are banned; systems deemed 'high-risk' are subject to 'essential requirements'; general purpose models are subject to transparency requirements, intellectual property protection, and 'systemic risk' mitigation; systems deemed 'limited risk' must fulfill transparency and accountability requirements; and systems with 'minimal risk' are largely exempt from regulation. The Act takes a cross-sectoral approach to its categorization of AI systems' risk levels, meaning that the level of risk is also determined by which sector systems are used in–for example, healthcare or education.

The Act demonstrates an implicit concern for human autonomy in its risk-based approach to regulating AI. In surveying the list of prohibited AI practices laid out in Article 5, we can see a clear intent on behalf of regulators to protect fundamental rights and human autonomy against potential harms caused by AI systems.[1] The list includes AI systems using subliminal techniques beyond an individual's conscious awareness or purposefully manipulative or deceptive techniques intended to influence or manipulate individuals' behavior. One of the examples the Commission gave in a briefing on this prohibition was a potential system in which "[a]n inaudible sound [played] in

[1] Crucially, there are a number of exceptions to these prohibitions, which will be discussed below.

truck drivers' cabins to push them to drive longer than healthy and safe [where] AI is used to find the frequency maximising this effect on drivers" (Mazzini 2021). Another example of banned AI systems are those that exploit individuals who are vulnerable due to their age, disability, or socio-economic situation, with the objective or effect of distorting or influencing the individual's behavior. Concern for human autonomy is evident in the explicit prohibition of these systems for their capacity to impair or override an individual's ability to take informed, autonomous decisions free from manipulation or deception. Also banned are systems used to predict a person's risk of committing a crime without human involvement; social scoring AI systems, which classify people based on their behavior, economic status and/or personal characteristics[2]; and systems that categorize individuals based on their biometric data in order to infer their race, political leanings, trade union membership, religious or philosophical beliefs, sex life or sexual orientation (Art. 5). Such systems are prohibited on the grounds that they may violate the rights to dignity and non-discrimination and the values of equality and justice. These prohibitions are foundational to Taylor's understanding of respect as the politics of universalism, in which every person must receive free and equal treatment regardless of their personal characteristics (Taylor 1992). Other AI applications that the Act prohibits are facial recognition systems using untargeted scraping of facial images from the internet or surveillance footage; emotion-recognition systems used in the workplace or in education; and 'real-time' remote biometric identification (RBI) systems in publicly accessible spaces. The protection of human autonomy is of utmost concern in the prohibition of these practices as well. For example, remote biometric identification systems are prohibited explicitly because they allow for "the identification of natural persons *without their active involvement*, typically at a distance [...]" (my emphasis added, Rec. 17).

In addition to the list of prohibitions for unacceptable-risk AI systems, the list of high-risk AI systems in Annex III also demonstrates implicit and explicit concern for fundamental rights and human autonomy. Many of the systems listed under the category high-risk are used in the public sector. High-risk systems used in migration, asylum and border control, for example, are designated as such for their risk to fundamental rights including free movement, non-discrimination, and protection of private life—all of which are not only important rights in and off themselves, but can also be understood as essential

[2] Although only if it leads to detrimental or unfavourable treatment in social contexts that are either unrelated to the contexts in which the data was originally collected, or that is unjustified or disproportionate. Article 5(1)(c) of the AI Act.

to the realization of human autonomy (Rec. 60). AI systems used in law enforcement are designated as high-risk for similar reasons: principally, for their potential impact on human liberty as well as to the rights of effective remedy and a fair trial (Rec. 59). AI systems used in education, employment, and public services are high-risk for their potential to impact rights including non-discrimination (Rec. 56–58). Recital 58 explicitly states that because public services are 'necessary for people to fully participate in society' (i.e. enjoy their full rational autonomy as individuals), AI systems used in the provision of these services shall be designated as high-risk. Recital 61 explicitly intends to enable the rational autonomy of judges by mandating that, while AI tools can support judicial decision-making, the final decision 'must remain a human-driven activity'.

So, in summary, in the Act's designation of risk categories to different AI systems, the protection of the development and exercise of autonomy appears to be the central concern.

2.2 Human Oversight Provisions

One area where the AI Act demonstrates explicit concern for human autonomy is through its provisions on the protection of human oversight. Article 14 states that high-risk AI systems must be designed and developed in such a way 'that they can be effectively overseen by natural persons'. In other words, in the deployment of these systems, there must be a human-in-the-loop to avoid unchecked algorithmic decision-making and ensure that humans have the information and ability to intervene if necessary. The enforcement of this provision mandates that, before putting a high-risk system on the market, for example for use in hiring, education, law enforcement, healthcare or critical infrastructure, providers must prove that the decisions taken by the systems will be monitored by humans who have the power to override or intervene when necessary. In order to meet this requirement, providers of high-risk systems are subject to explainability and transparency obligations. The Act requires that the operation of high-risk AI systems is transparent to the extent that its outputs can be understood, interpreted and predicted by the user (Art. 13). Individuals in charge of human oversight must be able to understand the capacities and limitations of the AI system, be aware of over-reliance (i.e. automation bias), be able to correctly interpret the system's output, and to ultimately decide, in any particular situation whether or not to use the high-risk system or to disregard, override, or reverse its output or interrupt its operation (Art. 14(4)). Taken together, these provisions demonstrate an explicit intent by regulators to prevent AI-systems from overriding human autonomy.

2.3 Transparency Obligations

The AI Act also demonstrates an intent to respect through its transparency provisions. Article 50 of the Act sets out specific transparency obligations for AI systems that interact with people to prevent manipulation, deception and loss of human autonomy. In order to ensure that users can make conscious decisions about their engagement with AI systems, companies must inform users when they interact with an AI system 'unless this is obvious from the point of view of a natural person who is reasonably well-informed, observant and circumspect' (Art. 50(1)). Users must also be informed if the AI system uses emotion recognition or biometric categorization. AI systems that generate or manipulate text, image, audio and video content (such as deepfakes) are required to clearly label their outputs as artificially generated in order to prevent deception. This requirement is limited when the content is 'evidently' artistic or satirical.

2.4 Compliance

Finally, the Act demonstrates an intent to respect individuals by enforcing compliance measures for companies and deployers of high-risk AI systems in order to hold them accountable for developing and deploying AI systems that ensure protection of human autonomy. These include risk management and documentation requirements, as well as fines for violations of up to 35 million euro for noncompliance. To monitor compliance and enforce penalties, the Act designates the establishment of an EU AI Office (Correa, 2025).

3 Respect in Practice

Section 2 above showed that the AI Act demonstrates an intent to respect individuals' autonomy through various prohibitions and provisions, including human oversight measures, transparency obligations, and compliance enforcement. While this intention is laudable in theory, however, I argue that in practice the Act falls short in delivering on its aim. The following section will critically analyze the Act's effectiveness in promoting self-respect through enabling human autonomy. It will explore as a first step how the approach taken by regulators impacts the Act's effectiveness in promoting respect, and as a second step how the Act's intentions to respect are undermined by its weak enforcement mechanisms and numerous loopholes.

3.1 Rights-Based Vs. Risk-Based Approaches

There are, broadly speaking, two possible legal approaches that can be taken in the development of digital policy: the risk-based approach and the rights-based approach. Rights-based approaches aim to empower individuals by granting them individual rights, which can include fundamental rights such as privacy or freedom of expression, or concrete rights such as the right to information or erasure (Leufer et al. 2024). Rights-based approaches thus focus "on the harm to individuals and on legal measure[s] empowering their initiatives" (Sartor and Lagioia 2020, p. 66). We can see immediately how this approach, which recognizes individuals as bearers of rights in all scenarios and emphasizes the empowerment of individuals, enables human autonomy and individual self-development. In other words, rights-based approaches clearly tie into the notion of recognition as respect. The EU's General Data Protection Regulation (GDPR) is a recent example of rights-based digital policy. Under its framework "individuals can exercise their personal data rights regardless of the perceived gravity of the use case[...] Hence, a more foundational and universally applicable level of protection is extended to individuals under a rights-based approach" (Leufer et al. 2024, p. 161).

Conversely, risk-based approaches do not grant individual rights but rather focus on reducing and preventing harm through "appropriate technical and organisational measures" (Gellert 2021, p. 20). Rather than granting individuals concrete rights, risk-based approaches focus on harm detection and mitigation through enforcing various accountability measures, required documentation procedures and assessment tools.

While the AI Act includes rights-based elements, its approach is primarily risk-based. Article 1 of the Act clearly lays out the order of its priorities: 'to improve the functioning of the internal market and promote the uptake of human-centric and trustworthy artificial intelligence (AI), while ensuring a high level of protection of health, safety, fundamental rights' (Art. 1(1)). Indeed, the Act is first and foremost a product safety regulation. It is based on Article 114 of the Treaty on the Functioning of the European Union (TFEU), which concerns the harmonization of national laws to reduce trade barriers and improve the functioning of the internal market. The Act takes a utilitarian approach to product safety, seeking to minimize risk while maximizing utility. It categorizes AI systems according to their determined level of risk and mandates subsequent mitigation strategies through various procedures and technological fixes to reduce and prevent instances of harm (Veale and Borgesius 2021). As such, fundamental rights protections are considered primarily as they relate to grave instances of harm from high-risk AI systems as outlined in Annex III—which section 2 of this chapter also explained.

3.2 Designating and Downgrading Risk

This mixed regulatory approach has significant consequences for the AI Act's effectiveness in promoting human autonomy. First, the risk-based approach relies on the regulation of AI systems according to a defined list of high-risk applications. As such, the designation of a given system as high-risk becomes a highly contentious political issue regarding who has the power to decide what constitutes acceptable and unacceptable levels of risk. In the development of technology policy, this normative decision-making power rests disproportionately with Big Tech companies, which enjoy outsized financial and personnel resources in relation to other lobbying players such as civil society groups, and upon which decision-makers often rely for technology-specific information (Floridi 2020). The creation of the AI Act sparked the most aggressive lobbying campaign in the history of the EU, with Big Tech companies launching unprecedented campaigns to push for, among other issues, the exclusion of general purpose models from being designated as inherently high-risk (Smuha 2023). This issue was so contentious that it led to Sam Altman threatening that OpenAI could decide to "cease operating" in Europe if it deemed itself unable to comply with the regulation (Perrigo 2023). The question of what constitutes a genuine risk is already shaping up to be one of the biggest battlefields of the AI Act's implementation through the standard setting process (discussed below in Sect. 3.3.2.2). Recent reports showed that Big Tech succeeded in watering down key provisions of the Code of Practice on General Purpose AI, which sets out standards for model providers to comply with the Act (Kroet 2025).

Second, the list of high-risk AI systems established in Annex III was intended to provide legal certainty, but this is undermined by the possibility for these systems to be downgraded to limited risk. The Act enables providers of high-risk AI systems listed in Annex III to self-assess whether their system actually poses a significant risk to health, safety, and fundamental rights. Article 6 of the Act states that, with the exception of profiling systems, a high-risk system included in the list in Annex III 'shall not be considered to be high-risk where it does not pose a significant risk of harm to the health, safety or fundamental rights of natural persons, including by not materially influencing the outcome of decision making' (Art. 6(3)). This includes, for example, AI systems intended to perform narrow procedural tasks and improve upon results of previously completed human activities. Should a provider consider that an AI system listed in Annex III is not high-risk, they must conduct a fundamental rights impact assessment before putting the system on the market or into service, register it with the appropriate national authority, and provide documentation of their risk assessment 'upon request of national competent authorities' (Art. 6(4)). Having met these documentation requirements, there is a presumption of conformity and the system can be downgraded to limited-risk. This critical procedure thus effectively comes down to corporate self-regulation.

3.3 Enforcement Effectiveness

For high-risk systems that are not downgraded to limited risk, the extent to which individuals' fundamental rights are protected and autonomy enabled ultimately depends on the on-the-ground effectiveness of the associated oversight and enforcement mechanisms. As outlined in the previous section, the bulk of the AI Act's regulatory muscle comes through its mandating of risk management systems and quality management systems for high-risk applications, as well as associated fines. To what extent are these mechanisms poised to promote human autonomy in practice?

3.3.1 Fines

The AI Act specifies fines based on the severity of the violation, with the highest fines, up to €35 million or 7% of worldwide annual turnover (whichever is higher), reserved for non-compliance with the prohibited practices outlined in Article 5 (Art. 99). However, it is unclear how strong of a deterrent penalties and fines will constitute in practice, particularly for large AI providers. While the amounts designated in the Act for noncompliance are by industry standards very high, they are, as Nathalie Smuha has argued, poised to be relatively ineffective against "companies whose annual revenues are higher than the GDP of several countries" (Smuha 2023, p. 11).

3.3.2 Compliance

If a provider determines that an AI system does constitute a high risk to EU citizens' safety and fundamental rights, it is up to the discretion of the provider how they wish to prove their compliance with the associated requirements. They can choose to rely on self-interpretation of compliance, on a Conformity Assessment Body, or on standardizations. As there is no obligation for providers of high-risk AI systems to use a Conformity Assessment Body, except in the case of biometric systems (Article 43), self-certification and standardization are the most common routes. These two routes will be analyzed below.

Self Certification

As long as providers of high-risk AI systems submit a declaration of conformity with the associated requirements (registration, risk management, technical documentation, etc.), they can put their system on the market without inspection, approval or licensing. This means that there is no independent, third-party oversight for high-risk AI systems before they enter the market. Rather, the role of public regulators is largely limited to ex post oversight: providers are required to put in place a risk-management monitoring system once the AI system is on the market (& Yeung, 2025). But this leaves European citizens vulnerable to harm in the meantime, as "several studies have found that products that have been self-certified by producers are considerably more likely to fail to meet the certified standard" (Smuha and Yeung 2025, p. 28).

The effectiveness of oversight for general purpose AI (GPAI) models with 'systemic risk' is similarly dubious. The legislator opted to express the criteria for qualifying GPAI models as having 'systemic risk' in terms of the threshold pertaining to the size of the data on which the models are trained. This threshold is not only arbitrary but it also means that smaller but more potent models could fall outside the scope of the Act (Hacker 2023). In addition to the requirements for all GPAI model providers for transparency and technical documentation and compliance with IP law, where GPAI models are deemed to pose a 'systemic risk' to fundamental rights, providers are additionally required to perform model evaluations and adversarial testing, assess and mitigate systemic risks, report serious incidents, and ensure an adequate level of cybersecurity (Art. 55(1)). Crucially, however, these audits and evaluations can be performed by providers of systemic risk GPAI models themselves and do not require external independent third-party audits (Smuha and Yeung 2025).

Standardization

Providers of high-risk AI systems can also choose to have the system standardized by one of the European Standardization Organisations (ESOs), including CEN (European Committee for Standardisation) and CENELEC (European Committee for Electrotechnical Standardisation). These organizations have been tasked with adopting standards relating to the mandate of the legislation. Providers of AI systems can then follow this standard instead of interpreting the AI Act's essential requirements themselves; if the standard is met, providers enjoy a presumption of conformity and their compliance burden is lessened (Art. 40). All of this means that, should an individual's rights be harmed by such a system, any subsequent accusations of non-conformity would be very difficult to prove.[3]

Because the standardization route is cheaper and a safer bet for providers, legal scholars have argued that this is where the "real rule-making" in the AI Act will occur; and as such, it could constitute an "unlawful delegation of the Commission's rulemaking power to private bodies" (Veale and Borgesius 2021, p. 105). Indeed, this standardization process is opaque and undemocratic. It is unclear, for example, how one defines in technical requirements what constitutes a risk to fundamental rights. As legal scholars Nathalie Smuha & Karen Yeung have pointed out, "technical standardisation processes may appear natural as they focus on mundane technical tasks, conducted in a highly specialised vernacular, yet these activities are in fact highly political" (Smuha and Yeung 2025, p. 30). Because these standards are not made publicly available due to copyright, civil society and the general public are not privy to how these standards are set. While civil society groups were invited to participate in the standard-setting process, relevant stakeholder organizations often face a lack of resources and expertise necessary to navigate the arcane private standardization processes (Veale and Borgesius 2021). All of this means that the communities most likely to suffer from adverse consequences of AI should its risks turn into rights violations are least likely to have their voices heard in this process.

[3] Under the AI Act, EU citizens have the right to lodge a complaint with a market surveillance authority and the right to explanation of individual decisions. These rights are, however, only for high-risk systems which affect an individual in a significant manner, and they are also subject to restrictions. Article 85 and Recital 170 of the AI Act.

In summary, in leaving oversight primarily to industry self-regulation without robust enforcement mechanisms, the Act leaves the fundamental rights of individuals vulnerable to harm. As such, the Act cannot be said to effectively promote the development of individuals' self-respect.

3.4 High-Risk Systems Used by the State

This lack of effective corporate oversight is particularly concerning considering the implications for the use of high-risk AI systems by the state. As outlined in Sect. 2.1, the majority of the high-risk systems listed in Annex III are designated as such for their use by various public authorities, including the provision of public services, law enforcement, and migration and border control. Yet in its operations and provision of services, these public authorities very often rely on private companies to provide the AI systems that they use. This set-up can lead to state dependence on private AI providers and information asymmetry, which in turn has a number of adverse implications for enforcing human oversight and ensuring justice.

The AI Act mandates that providers of high-risk AI systems build human oversight into the system for use by deployers–in this case, the state (Art. 14). As such, the responsibility for human oversight rests primarily with the companies providing the AI system instead of the state authority using it. This creates the potential for a situation in which a government actor is using an AI system for which it does not fully understand (nor have power over) how the outputs are created. As a result, individuals' fundamental rights could be left vulnerable to harm in their relations with the state. This was the case in the Dutch child welfare scandal, in which an algorithmic decision-making system deployed by the Dutch government to detect fraudulent child benefit applications relied on racial profiling in its risk assessments. The use of this system by the state resulted in tens of thousands of parents and care givers primarily from low-income families and migrant backgrounds being falsely accused of committing fraud, causing severe financial and psychological harm to individuals and communities. This algorithmic discrimination was made possible due to the lack of effective oversight by the state over the AI system it was deploying: when an individual was flagged as a fraud risk, a civil servant was required to conduct a manual review but was given no information as to why the system had generated a higher-risk score (Amnesty International 2021). With its significant shortfalls in enforcement and oversight mechanisms, it is unclear how the AI Act will effectively protect citizens from such AI-driven fundamental rights violations in the future.

Following pressure from the European Parliament, the legislator included a provision that high-risk AI deployers that are bodies governed by public law or are private entities providing public services must also carry out a fundamental rights impact assessment before putting a high-risk system into use (Art. 27). However, as Smuha & Yeung argue, "the fact that an 'automated tool' will be provided to facilitate compliance with this obligation 'in a simplified manner' suggests that the regulation of these risks is likely to descend into a formalistic box-ticking exercise in which formal documentation takes precedence over its substantive content and real-world effects" (Smuha and Yeung 2025, p. 17). In short, the fundamental rights impact assessment is unlikely to do much for the protection of fundamental rights and enabling of human autonomy within the framework of weak enforcement and oversight mechanisms.

3.5 Loopholes and Exemptions

The AI Act's effectiveness in enabling human autonomy is also undermined by its many legal loopholes and exemptions, particularly regarding the use of high-risk systems by law enforcement and migration authorities. These include exemptions from human oversight and transparency requirements, loopholes for specific high-risk technologies, and general exemptions for national security.

3.5.1 Transparency Exemptions

According to Article 50(1) of the Act, transparency obligations for providers and deployers of high-risk AI systems do not apply to AI systems 'authorised by law to detect, prevent, investigate or prosecute criminal offences' (Art. 50(1)). This means that individuals will no longer have the right to be informed when they are subject to decisions informed by an AI system,[4] when they are interacting with an AI system or being subjected to an emotion recognition or biometric categorization system, when they are being exposed to AI-generated content, to have explanations about algorithmic decisions, or to know that they are being subjected to the testing of AI systems. Through this exemption, the Act *inhibits* rather than enables individuals' rational autonomy as it provides the conditions for high-risk AI systems to be used against them without their knowledge—including systems in potential violation of

[4] However, individuals will have rights under the GDPR to be informed of any such decisions.

their fundamental rights, such as dignity, non-discrimination, equality and justice.

3.5.2 Oversight Exemptions

Recital 73 of the Act states that, 'given the specificities of the areas of law enforcement, migration, border control and asylum' the human oversight provisions listed in Article 14 'should not apply where Union or national law considers the application of that requirement to be disproportionate'. While the use of AI systems by policing and migration authorities will be overseen by data protection authorities, these bodies are severely under-funded and lacking in resources. Moreover, the Act appears to grant policing and migration authorities the ability to restrict or prevent supervisory authorities from exchanging information regarding their use of AI systems, as well as control over supervising bodies' access to their technical documentation (Statewatch n.d.). All of this means that meaningful supervision of and control over the use of AI systems for law enforcement and border control will be very difficult. This exemption effectively inhibits individuals' autonomy by preventing them from having any information regarding the use of intrusive and invasive AI systems by these authorities.

3.5.3 'Real-Time' Remote Biometric Identification Loophole

Perhaps the most controversial of these loopholes is for 'real-time' remote biometric identification (RBI). As outlined in Sect. 2.1 of this chapter, the AI Act bans the use of 'real-time' remote biometric identification by law enforcement in publicly accessible spaces; however, this prohibition includes a number of exceptions. These are: the targeted search for specific victims of crime including abduction and human trafficking, and for missing persons; the prevention of a specific, substantial and imminent threat to the life or physical safety of natural persons, including terrorist attacks; and the localization or identification of a person suspected of having committed a criminal offence (Art. 5). Before using these systems in publicly available spaces, law enforcement authorities are required to carry out a fundamental rights impact assessment (Article 27) and to register the system in the EU database (Article 49); however, this registration can be done post-use in 'duly justified cases of urgency'. The effectiveness of fundamental rights impact assessments, which are likely to become box-checking exercises, has already been questioned.

Moreover, the prohibition applies only to 'real-time' systems that capture, compare, and identify 'instantaneously, near-instantaneously or in any event without a significant delay'. This excludes 'post' systems for which 'the biometric data has already been captured and the comparison and identification occur only after a significant delay' (Rec. 17), and which include, for example, the use of biometric systems to identify individuals at protests after the fact, as well as systems that categorize individuals biometrically. Online spaces also fall outside the scope of this prohibition, meaning that live biometric identification can be used, for example, on video streams (Veale and Borgesius 2021). Lastly, the prohibition excludes the use of RBI for non-law enforcement purposes. It does not apply for use in border control, nor for use by private companies. This means that, under the AI Act, these systems can be used for practices as wide ranging as border control, to crowd control or public health.[5] So extensive are these loopholes that legal scholars have questioned whether this prohibition can even be considered as such in practice (Giannini and Tas 2024).

Recital 32 of the Act outlines the risks that 'real-time' RBI poses to fundamental rights:

> "The use of AI systems for 'real-time' remote biometric identification of natural persons in publicly accessible spaces for the purpose of law enforcement is *particularly intrusive to the rights and freedoms* of the concerned persons, to the extent that it may *affect the private life of a large part of the population, evoke a feeling of constant surveillance and indirectly dissuade the exercise of the freedom of assembly and other fundamental rights.* Technical inaccuracies of AI systems intended for the remote biometric identification of natural persons can lead to *biased results* and entail *discriminatory effects.* Such possible biased results and discriminatory effects are particularly relevant with regard to age, ethnicity, race, sex or disabilities. In addition, the immediacy of the impact and the limited opportunities for further checks or corrections in relation to the use of such systems operating in real-time carry *heightened risks for the rights and freedoms of the persons concerned* in the context of, or impacted by, law enforcement activities."

Clearly, 'real-time' RBI systems involve severe and manifold threats to fundamental rights, including freedom of assembly and freedom to a private life. In allowing for their use by law enforcement, the AI Act not only fails to protect individuals' rights and freedoms, but effectively paves the way for the use of

[5] Although these uses would fall under the GDPR.

highly intrusive and invasive systems with grave implications for individuals' ability to enjoy their full human autonomy.

3.5.4 National Security Exemption

Member States fought hard to ensure that their competences in national security remained untouched, and the AI Act grants this. The Act states that its provisions do not apply to AI systems developed or used 'solely for military, defence, or national security purposes, regardless of the type of entity carrying out those activities' (Article 2). In effect, this means that, in one of the most potentially intrusive areas for which AI could be used, protections for individuals are very weak (Smuha and Yeung 2025).

In effect, this loophole enables Member States to develop and use high-risk AI systems if they self-authorize this use under the Regulation's blanket exemption for national security. In short, what constitutes national security, and, in this case, legitimate grounds for deploying high-risk technologies, is a political question determined by Member States themselves. As such, "climate demonstrations or political protests, for instance, could now be freely targeted with AI-powered surveillance if police have national security concerns" (Maggiore et al. 2025). Similar concerns arise for border control, which is increasingly being framed in European policy discourse as a national security issue. If Member States evoke the national security exemption in migration and asylum control, refugees and migrants entering the EU will be left with even fewer protections from the use of intrusive AI systems than they are currently afforded under the regulation.

Legal scholars argue that this loophole "paves the way for Member States to install and equip public places with facial recognition cameras which can then be configured for the purposes of remote biometric identification if the exceptional circumstances are met, thus expanding the possibility of function creep and the abuse of law enforcement authority (Smuha and Yeung 2025, p. 12). This exemption has been also been heavily criticized by civil society groups, who warn that it paves the way for the use of mass surveillance systems in the EU (ReclaimYourFace n.d.).

4 Who Should Be the Subject of Respect in AI Policy?

The above discussion of the ways in which the AI Act fails to enable the development and exercise of autonomy mostly showed how this AI policy disrespects all members of society equally. However, since recognition as respect has to do with the equal treatment–or "the politics of universalism", as Charles Taylor called it (Taylor 1992)–we should also ask whether there are any specific demographics that are misrecognized, because they are left out of consideration, in the AI Act. In other words, in addition to analyzing the extent to which AI policy enables autonomy, considering recognition as respect in the context of AI policy also raises questions like: Who is protected by this law, and who should be protected by it? Or, who does the Act recognize as bearers of rights, and who does it fail to recognize? Article 2 limits the scope of the Act to 'affected persons that are located in the Union' (Art. 2). As such, the protections that it does offer to individuals are limited to those within its borders. But what about individuals who are not located within the Union, but against whom AI systems developed or deployed by European companies or institutions are used?

4.1 Refugees and Migrants

Refugees and migrants seeking status in the EU are left highly vulnerable to harm by the Act's loopholes for migration control and border security. People seeking to enter the EU could be subject to the use of a number of high-risk AI systems, including predictive analytics, profiling systems, emotion recognition or 'lie detector' systems, and Remote Biometric Identification (RBI). These systems can be used to curtail migration by surveilling borders and predicting patterns of 'irregular migration', to assess whether people on the move pose a 'risk' of unlawful activity or security threats, and to detect 'suspicious' people with behavioural analytics and the inference of emotion from biometric data. As such, the use of these systems has the potential to violate a number of fundamental rights tied explicitly to the positive development of human autonomy, including the right to seek asylum and to *non-refoulement*, the right to equality and non-discrimination, the presumption of innocence, and human dignity. Through its extensive border security loopholes, the Act fails to recognize refugees and migrants seeking status in the EU as bearers of rights who ought to be treated equally to other members of society. Moreover, due to the limit of the Regulations scope to individuals inside the EU, refugees and migrants located outside the EU have, under the AI Act, no right to obtain 'clear and meaningful explanations' of an AI system's role in a decision that impacts their health, safety or fundamental rights.[6]

4.2 Individuals Located Outside the EU

The Act also fails to recognize as bearers of rights individuals located outside the EU whose rights could be harmed by AI systems developed by European companies. Another glaring example of the Act's failure to recognize individuals outside the EU as equal bearers of rights is evident in the loophole for 'real-time' RBI systems that are developed in the EU and used elsewhere. Unlike the rest of the prohibited practices listed in Article 5, which prohibit not only the use of the systems but also their placement on the market, the provision for remote biometric identification is prohibited only in use, meaning that such biometric systems are allowed to be placed on the market. In effect, this means that EU vendors can sell biometric systems which would be illegal to use in the EU to oppressive regimes in third countries, for use against their citizens (Veale and Borgesius 2021). This is not a hypothetical risk but is already happening. Amnesty International has uncovered multiple cases of such digital surveillance exports, including the French firm Idemia/Morpho selling facial recognition technology to the Shanghai Public Security Bureau, and the Dutch firm Moldus selling the facial expression analysis tool 'FaceReader' to the Chinese Ministry of Public Security (Amnesty International 2020).

The Act also fails to recognize individuals located outside the EU whose rights have been violated through harmful data collection practices, which European AI companies subsequently benefit from. Article 10(3) of the Act states that 'training, validation and testing data sets shall be relevant, sufficiently representative, and to the best extent possible, free of errors and complete in view of the intended purpose'. In failing to sufficiently address how the GDPR's requirements for collecting and processing data interact with its provisions, the AI Act leaves open the question of whether data that has been collected in violation of the rights of people located outside the EU would be deemed appropriate by the Regulation. As such, it fails to protect the rights of individuals living in regimes with less extensive privacy rights (Smuha et al. 2021).

[6] The right to explanations may be available under data protection law such as the GDPR or from the right to effective remedy in the EU Charter of Fundamental Rights.

4.3 AI Workers

Another group of stakeholders who the Act also fails to recognize are the individuals impacted by the development of AI systems along the production chain, in particular those whose labor contributes to the development and use of AI systems. These include, for instance, miners who dig up the rare minerals needed for the hardware in datacenters and data annotators who give meaning to the data that AI systems are trained on. In many cases, these workers are subjected to various labour injustices, including poor working conditions, unlivable wages, and a lack of job security. Without these workers, EU companies, authorities, and citizens would not be able to use and benefit from AI. Yet these individuals are entirely absent from the AI Act. This neglect represents a failure to recognize not only their contributions to the technologies in question, but also a failure on behalf of the EU to acknowledge the equal worth of AI workers. This failure to confer respect to AI workers contributes to their hiddenness and reinforces the precarity of AI labor practices (Muldoon et al. 2024).

4.4 Policy and Normativity

These loopholes and exclusions represent political choices made by policy makers. It is therefore worth interrogating the normative considerations underpinning these choices. Why are certain technologies that are more intrusive and harmful to fundamental rights allowed to be used against certain individuals and groups and not others? Why is it that refugees and migrants need to give up more of their basic liberties at the border than EU citizens do? Why are the same systems that are rightfully deemed unacceptable risks to fundamental rights in the EU allowed to be developed by European companies and used against non-Europeans in other countries? Are the fundamental rights of non-EU citizens somehow lesser than those of EU citizens?

In his conceptualization of recognition as respect, Honneth focuses on institutional rights conveyed within national communities. Yet, recognition theory can apply transnationally, particularly when considered in light of universal principles of fundamental rights that are governed by international institutions (see also Schweiger 2012). In failing to protect the rights of individuals located outside of the EU against potential harms posed by high-risk AI systems, the AI Act fails to recognize these individuals as bearers of rights, who ought to be treated fairly and equally as other members of society. As such, it fails to promote their exercise of autonomy and development of a positive relation-to-self.

5 Conclusion

Ultimately, the AI Act cannot be said to respect individuals in practice because it does not effectively protect their fundamental rights and thus enable them to enjoy their full human autonomy and develop positive relations-to-self. The lack of effective fundamental rights protection that characterizes the Regulation is a result of the risk-based approach taken by the legislator. While the Act includes elements of a rights-based approach, it is, primarily, a product safety legislation concerned with reducing trade barriers and improving the functioning of the internal market. It is concerned with fundamental rights to the extent that their protection does not constitute a disproportionate burden on AI developers and a challenge to the EU's AI competitiveness. As such, it demonstrates an intent to promote the rational autonomy of EU citizens through transparency and human oversight requirements, but it is unlikely to achieve this in practice due to the ineffectiveness of its enforcement mechanisms.

Moreover, the Act cannot be said to respect individuals because it does not meet the basic condition of the politics of universalism: that is, recognizing individuals as bearers of rights who ought to be treated equally to other members of society, regardless of which national community they belong to or where they are located. Rather, the Act prioritizes the rights of individuals located within the EU, while leaving those outside of it vulnerable to harm by the use of high-risk AI systems deployed by EU agencies or developed by European companies.

Ultimately, the most meaningful protection of fundamental rights and promotion of autonomy will come from other areas, primarily rights-based legislation. The Act states that it should not affect 'existing Union law, in particular on data protection, consumer protection, fundamental rights, employment, and protection of workers, and product safety' (Recital 9). It remains to be seen how the provisions of the Act that appear in tension with other EU laws will be implemented, and how these will be interpreted by the courts. The AI Act's shortfalls in fundamental rights protection could be mitigated by the EU Charter of Fundamental Rights, as well as by existing data protection law such as the GDPR. The EU Data Protection Supervisor, which was set up to enforce the provisions of the GDPR, will remain a key player in monitoring fundamental rights protection in EU agencies, including Frontex and Europol, particularly in their use of high-risk technologies such as biometric surveillance. The robustness of the GDPR in this regard is thanks in large part to its rights-based approach, which recognizes individuals as bearers of rights in all scenarios and enforces legal measures to empower their initiatives. The AI Act's failure to adopt a similar approach represents a missed opportunity in the recognition of individuals through AI policy.

References

Amnesty International. 2020. *Out of control: Failing EU Laws for digital surveillance export. Report. Amnesty international ltd.* Amnesty International Ltd. https://www.amnesty.de/sites/default/files/2020-09/Amnesty_Bericht_Digitales_Out_of_Control_SPERRFRIST_21.09.2020.01.01_Uhr_MESZ%20%281%29.pdf. Accessed 20 May 2025.

Amnesty International. 2021. *Xenophobic machines: Discrimination through unregulated use of algorithms in the Dutch childcare benefits scandal. Report. Amnesty international ltd.* Amnesty International Ltd. https://www.amnesty.org/en/documents/eur35/4686/2021/en/. Accessed 5 August 2025.

European Union. Regulation (EU) 2024/1689 of the European Parliament and of the Council of 13 June 2024 Laying Down Harmonised Rules on Artificial Intelligence (Artificial Intelligence Act) and Amending Regulations (EC) No 300/2008, (EU) No 167/2013, (EU) No 2018/858 and (EU) 2023/2854. Official Journal of the European Union, L 168, 12 June 2024, pp. 1–159. https://eur-lex.europa.eu/eli/reg/2024/1689/oj.

Floridi, Luciano. 2020. The fight for digital sovereignty: What it is, and why it matters, especially for the EU. *Philosophy & Technology* 33 (3): 369–378. https://doi.org/10.1007/s13347-020-00423-6.

Gellert, Raphaël. 2021. The role of the risk-based approach in the general data protection regulation and in the European Commission's proposed artificial intelligence act: Business as usual? *Journal of Ethics and Legal Technologies* 3 (2): 15–33. https://doi.org/10.14658/pupj-JELT-2021-2-2.

Giannini, Alice, and Sarah Tas. December 10, 2024. *AI Act and the Prohibition of Real-Time Biometric Identification. Verfassungsblog*,. https://doi.org/10.59704/a15aa6e58151c853. Accessed 5 May 2025.

Hacker, Philipp. December 13, 2023. What's Missing from the EU AI Act: Addressing the Four Key Challenges of Large Language Models, VerfassungsBlog, https://verfassungsblog.de/whatsmissing-from-the-eu-ai-act/. Accessed 10 April 2025.

High-Level Expert Group on AI (AI HLEG). April, 2019. *Ethics guidelines for trustworthy AI.* https://ec.europa.eu/digital-single-market/en/news/ethics-guidelines-trustworthy-ai. Accessed 20 May 2025.

Kroet, Cynthia. May 2, 2025. Big Tech Watered Down AI Code of Practice: Report. *Euronews*, https://www.euronews.com/next/2025/04/30/big-tech-watered-down-ai-code-of-practice-report. Accessed 17 May 2025.

Leufer, Daniel, Fanny Hidvegi, and Alessia Zornetter. 2024. The pitfalls of the European Union's risk-based approach to digital rulemaking. *UCLA Law Review*. https://www.uclalawreview.org/the-pitfalls-of-the-european-unions-risk-based-approach-to-digital-rulemaking/. Accessed 2 May 2025.

Mazzini, Gabriele. 10 May 2021. 'A European Strategy for Artificial Intelligence' *(2nd ELLIS Workshop in Human-Centric Machine Learning (YouTube recording)*) https://youtu.be/OZtuVKW qhl0?t=10346. Accessed 10 June 2025.

Maggiore, Maria, Leila Minano & Harald Schumann. January 22, 2025. France Spearheaded Successful Effort to Dilute EU AI Regulation. *EUobserver*, https://euobserver.com/digital/ardc3193c4. Accessed 23 January 2025.

Muldoon, James, Mark Graham, and Callum Cant. 2024. *Feeding the machine. The hidden human labour powering AI*. Canongate.

Perrigo, Billy. June 20, 2023. OpenAI Lobbied E.U. to Water Down AI Regulation, Time, accessed June 14, 2025, https://time.com/6288245/openai-eu-lobbying-ai-act/.https://time.com/6288245/openai-eu-lobbying-ai-act/. Accessed 8 May 2025.

ReclaimYourFace. n.d. Members of the European Parliament: Will You Stand up for Our Rights? - Reclaim Your Face. Reclaim Your Face, https://reclaimyourface.eu/meps-will-you-stand-up-for-our-rights-and-ban-bms/. Accessed 21 May 2025.

Sartor, Giovanni, and Francesca Lagioia. 2020. Panel for the Future of Science and Technology, European Parliamentary Research Service, and Scientific Foresight Unit (STOA). "The Impact of the General Data Protection Regulation (GDPR) on Artificial Intelligence." Report. *STOA*, https://www.europarl.europa.eu/RegData/etudes/STUD/2020/641530/EPRS_STU(2020)641530_EN.pdf. Accessed 27 April 2025.

Schweiger, Gottfried. 2012. Globalizing recognition: Global justice and the dialectic of recognition. *Public Reason* 4.

Smuha, Nathalie A. July 1, 2023. Digital Sovereignty in the European Union: Five Challenges from a Normative Perspective, SSRN Scholarly Paper Rochester, NY, https://doi.org/10.2139/ssrn.4501591.

Smuha, Nathalie A., Emma Ahmed-Rengers, Adam Harkens, Wenlong Li, James MacLaren, Riccardo Piselli, and Karen Yeung. 2021. How the EU can achieve legally trustworthy AI: A response to the European Commission's proposal for an artificial intelligence act. *SSRN Electronic Journal*. https://doi.org/10.2139/ssrn.3899991.

Smuha, Nathalie, and Karen Yeung. 2025. The European Union's AI act: Beyond motherhood and apple pie? In *The Cambridge handbook of the law, ethics, policy and artificial intelligence*, edited by Nathalie Smuha, 228–258. Cambridge University Press. https://doi.org/10.1017/9781009367783.015. p. 25.

Statewatch. n.d. Automating Authority: 2. Cop Out: Security Exemptions in the Artificial Intelligence Act, https://www.statewatch.org/automating-authority-artificial-intelligence-in-european-police-and-border-regimes/2-cop-out-security-exemptions-in-the-artificial-intelligence-act/#_ftn22. Accessed 1 June 2025.

Taylor, Charles. 1992. The politics of recognition. In *Multiculturalism: Examining the politics of recognition*, edited by A. Gutmann. Cambridge: Princeton University Press.

Thompson, Simon. 2006. *The political theory of recognition: A critical introduction*. Polity.

Veale, Michael, and Frederik Zuiderveen Borgesius. 2021. Demystifying the draft EU artificial intelligence act — Analysing the good, the bad, and the unclear elements of the proposed approach. *Computer Law Review International* 22 (4): 97–112. https://doi.org/10.9785/cri-2021-220402.

Moral Recognition Relations Between Lifeforms and AI?

Christina Pinsdorf

Abstract This chapter re-examines moral recognition theory in the age of artificial intelligence. Building on Fichte's and Hegel's insight that self-consciousness and normativity arise through mutual acknowledgement between fully fledged persons and the integration of Michael Thompson's life-form ontology, as well as the extension of Philippa Foot's natural normativity, the analysis draws on a two-axis recognition typology—symmetry vs. asymmetry, reciprocity vs. non-reciprocity—developed in Pinsdorf (2016). This original typology is meant to capture all moral-recognition relations existent among human and non-human lifeforms.

However, with their increasingly pronounced occupation of the human's social lifeworld, LLMs, social chat-bots, and social robots challenge this typology and call for a reconsideration of whether and how recognition applies to non-biological entities, too. This raises questions concerning moral agency and patiency, whom or what qualifies as recognition giver and/or recognition receiver and the significance of authenticity and simulation. A properties-based analysis shows that current and foreseeable AI-systems are neither moral agents, nor moral patients: they lack autonomous practical reason, phenomenal (self-)consciousness, motivational attitudes and intrinsic goods.

C. Pinsdorf (✉)
Institute for Science and Ethics, University of Bonn, Bonn, Germany
e-mail: pinsdorf@uni-bonn.de

R. Waelen (ed.), *Normative Relations of Recognition in the Age of AI*, Philosophy and Ethics of AI 2, https://doi.org/10.1007/978-3-032-34028-3_9

Nevertheless, humans frequently experience AI interactions as recognizing due to anthropomorphic projection, design-induced social cues, and the opacity of machine reasoning. These encounters inspire a relational approach and instantiate pseudo-recognition: relations that mimic the grammar of acknowledgment but lack its normative substance. Although remaining illusionary, pseudo-recognition is ethically significant in two derivative ways: AI can function as recognition facilitator and provide recognition-like stimuli. Both roles generate indirect duties grounded in our obligations towards the human users who invest meaning and emotion in these artefacts. The analysis adheres to conceptual clarity of moral recognition theory between lifeforms while acknowledging the affective realities of human-AI interaction. It recommends a stance of respectful stewardship: safeguarding the relational goods people build around AI without attributing unwarranted moral status to the systems themselves.

Keywords Moral recognition theory • AI-lifeform-relations • Symmetry/asymmetry • Reciprocity/non-reciprocity • Pseudo-recognition • Properties approach • Relational approach

1 Introduction

In the tradition of moral recognition theory, the recognition relation exists only between fully fledged persons. Thinkers such as Fichte (1971 [1796]), Hegel (2006 [1807]), or currently Honneth (1992) and Ikäheimo (2009) conceive recognition as a process through which individuals both affirm and become affirmed in their status as person. The symmetry and reciprocity of moral recognition relations structure the ethical, social and legal lifeworld of fully-fledged persons. Although these mutual relations between moral agents form the core of moral recognition, they represent only one segment of all of the existing moral recognition relations. In philosophy of nature and ecological ethics, contemporary lifeform theory has convincingly revealed non-human living beings as worthy of moral consideration within the framework of asymmetric/non-reciprocal recognition relations (Pinsdorf 2016).

However, with the rapid advancement of AI-systems and their increasingly widespread socially embedded use, moral recognition relations as well their conditions are challenged to be reconsidered once again. Since the ontological status of AI-systems is disputed, but users report their strong emotional bonds, this paper explores the philosophical consequences of these developments and the related question: Can AI-systems be moral agents and/or moral patients. Can they be recognition givers and/or recognition receivers?

To address this question, the analysis first traces the conceptual genealogy of recognition (Sect. 2.1). Johann Gottlieb Fichte (1971 [1796]) famously argues that self-consciousness is impossible without reciprocal acknowledgment by another rational being; recognition is thus simultaneously epistemic and normative. G. W. F. Hegel (2006 [1807]) deepens the claim by showing how the *struggle for recognition* propels socio-historical development from brute encounters to institutionalized norms. Contemporary theorists such as Axel Honneth (1992), Ikäheimo and Laitinen (2007) and Pinsdorf (2016) unfold recognition as a triadic relation—X recognizes Y as Z—interweaving *cognition* (correctly *identifying* the other), *evaluation* (*attributing* normative status), and *motivation* (*acting* from a recognitive attitude).

A second argumentative strand relies on the advancement of Michael Thompson's lifeform concept and Philippa Foot's idea of natural normativity (Sect. 2.2). Lifeforms are teleologically structured wholes that fuse descriptive and normative aspects. From this vantage point, moral considerability is not exhausted by rationality or sentience. Every living being, simply by instantiating a lifeform, comes under natural normativity: it has a natural goodness (Foot 2001) *and* a natural beneficialness (Pinsdorf 2016). Recognizing a being therefore entails grasping both its form-relative standard of excellence (natural goodness) and the conditions conducive to its wellbeing (natural beneficialness). The present work exploits these insights to argue that flourishing constitutes the true minimal threshold of direct moral relevance.

Building on these conceptual resources, the study (Sect. 3) proposes a three-quadrant typology of moral recognition, generated by crossing the axes of symmetry/asymmetry and reciprocity/non-reciprocity (Pinsdorf 2016): (1) Symmetric-reciprocal recognition relations remain the classical paradigm: two moral agents, equal in dignity, mutually acknowledge one another's rights and duties. (2) Symmetric-non-reciprocal recognition relations capture cases such as parent–nursling or physician–comatose-patient interactions, where equal moral status coexists with an inability to reciprocate of one party. (3) Asymmetric-non-reciprocal recognition relations extend unidirectional moral concern to non-human lifeforms, whose moral standing is weighable rather than absolute.

Is this typology exhaustive, or does the accelerating diffusion of sophisticated AI-systems such as LLMs, social chat-bots, and social robots, threaten to unsettle it? Can AI be a giver or receiver of moral recognition? It is probed (Sect. 4) whether contemporary AI-systems can meaningfully occupy either pole of a moral-recognition relation. Recognition giving presupposes moral agency; recognition receiving presupposes moral patiency—it is demonstrated why and how present-day AI-systems despite their growing rational capacities, lack the necessary properties.

But even though AI-systems lack the properties grounding moral status—self-consciousness, sentience, or at least the capacity to flourish—their convincingly social behavior can still trigger powerful experiences. These encounters give rise to the notion of pseudo-recognition and the relational approach (Sect. 5), shifting the focus from "What intrinsic qualities make a being morally considerable?" to "How do our lived interactions confer moral weight?" The concept of pseudo-recognition makes experientially potent relations of indirect moral relevance comprehensible. Human-AI-relations generate indirect duties grounded in our obligations towards the human users who invest meaning and emotion in these artefacts. These relations may be ethically salient, not for the artefacts' sake but for the way we treat AI-systems may shape our own moral character as well as the moral character of society. It is concluded (Sect. 6) that the conceptual clarity of moral recognition theory needs to be asserted while the affective realities of human-AI interaction need to be acknowledged as pseudo-recognition. Pseudo-recognition and its long-term impact on human character and social norms raise a plentitude of ethical follow-up questions that need to be addressed by future research.

2 Historical and Conceptual Background

2.1 Recognition

Fichte's *Grundlage des Naturrechts nach Principien der Wissenschaftslehre* (Fichte 1971 [1796]) offers one of the earliest systematic accounts of recognition as a pre-condition for subjectivity. He contends that an individual cannot attain self-consciousness in isolation; awareness of oneself as a rational being arises only through the acknowledgment of another rational being. This mutual recognition is not merely epistemological but normative: it imposes a duty to respect the other as free and rational, thereby grounding legal and moral obligation. Crucially, for Fichte recognition is reciprocal and symmetric, because "neither can recognize the other unless both recognize each other: and neither can treat the other as a free being unless both treat each other as such" (Fichte 1971 [1796], p. 44 (own translation)). According to Fichte, the very 'human form'—indeed a mere 'face'—evokes a shared self-image that awakens a pre-rational desire for mutual recognition.

Hegel extends this insight into a broader dialectic of social development. In the *Phänomenologie des Geistes* (Hegel 2006 [1807]) recognition powers both the formation of self-consciousness and the course of history. The famous master–slave dialectic shows how the struggle for recognition becomes entangled with power, dependency, and existential risk. Recognition progresses from immediate encounters to institutionalized forms—family, civil society, and the state—where it is mediated by roles, rights, and norms. Thus, for Hegel recognition constitutes a structural dimension of freedom within ethical life (*Sittlichkeit*), rather than a mere interpersonal gesture.

Moral recognition can be schematized as a three-place relation: X (the recognition giver) recognizes Y (the recognition receiver) as Z (morally worthy of consideration in accordance with Y's moral status). The concept sharpens three inseparable dimensions of moral relations: the cognitive, the normative, and the motivational. *Cognitively*, recognizing (*an*erkennen) presupposes correct cognizing (*er*kennen),[1] one must first identify the entity appropriately. *Normatively*, one evaluates the entity's standing. *Motivationally*, one feels moved to act accordingly. Following contemporary recognition theorists like Honneth, Darwall, Ikäheimo and Laitinen (Honneth 1992; Darwall 2009; Ikäheimo and Laitinen 2007; Laitinen 2016) moral recognition is construed as the motivational readiness to take another being's standpoint into account.[2] Hence to recognize an entity x is to grasp it as x and to "simultaneously appreciate it as the bearer of a normative status in a motivationally effective way and reacting accordingly" (Laitinen 2009, p. 316 (own translation). See also Ikäheimo 2009). The morally relevant status at stake may be absolute (human dignity) or weighable (the ponderable worthiness of non-human living beings).

Contemporary debates—shaped by Fichtean or Hegelian lines of thought and pursued by figures such as Jürgen Habermas, Ludwig Siep, Axel Honneth, Charles Taylor, and Nancy Fraser (Siep 1979; Honneth 1992; Taylor 1992; Habermas 1997; Fraser 2003)—concur that persons *capable of reason and morality* are the only conceivable *givers of recognition*.[3] The capacity for reason may generally be understood as the ability to base thoughts and actions on

---

[1] The precise relationship between cognition and recognition, the direct connection between descriptivity and normativity, is elaborated in Pinsdorf (2016), Chap. VIII with regard to the integrative concept of the lifeform.

[2] The more a lifeform deviates from the human lifeform, the more difficult it is for humans to understand the standpoints of its specimens. Via the category of lifeform, it remains however always possible to approximate a specimen's standpoint. (Pinsdorf 2016).

[3] Laitinen departs from the dominant view by allowing animals to serve as potential recognizers (Laitinen 2009). Contemporary political and socio-philosophical theories of recognition typically enlarge the class of recognizers beyond individuals to include groups, institutions, and states. Yet, even on this broader

universally intelligible reasons, i.e., to make conscious and reflective decisions that go beyond mere instincts or habits. The capacity for morality may generally be understood as the ability to distinguish between good and evil, to recognize moral reasons as binding, and to align one's actions according to one's discernments.

By contrast, who or what can be a *recipient of recognition* remains contested. Within the Hegelian tradition recognition is anchored in two agents who reciprocally recognize each other as recognizers. Ikäheimo's contemporary account preserves this structural reciprocity as the core of any recognition relation (Ikäheimo 2009).

Demanding identical capacities in both recognition giver and recognition receiver is, however, implausible in many applied-ethics contexts and generates serious moral difficulties (Nussbaum 2006). If only individuals who themselves possess reason and moral agency can be recipients of recognition, many humans who lack those capacities—and who are therefore especially vulnerable—would be excluded. Yet newborns, comatose patients, and persons with severe intellectual disabilities stand in particular need of recognition by those who are fully endowed with reason and morality.

2.2 Lifeform

The understanding of the concept "lifeform" as used here goes back to Michael Thompson (2008). Drawing on Hegel, Thompson conceives of the lifeform as a logical category of human thought. Lifeforms are not mere empirical catch-all notions but analytic a priori structures that first render observational data intelligible. In this empirically rich sense, the lifeform unites the generality of an abstract concept with the particularity of concrete organisms—much like the Aristotelian *eidos* in which form and matter are fused (Aristotle 1988).

Descriptions of lifeforms consist of natural-historical teleological judgments that inextricably intertwine descriptive and normative elements. Anyone who asserts, in such a judgment, that "the cat has four legs" is invoking a general standard against which particular specimens are measured. Philippa Foot shows that these judgments are evaluative-normative because they register the criteria by which the success or failure of a life-process is to be assessed (Foot 2001). Natural-historical teleological judgments form a coherent web in which every statement about a living being refers to its form, its developmental trajectory, and its functional interconnections.

footing, every recognitive relation ultimately traces back to an association of individual persons endowed with the capacity of reason and morality.

Against this backdrop, Foot's concept of natural normativity can be expanded to encompass not only *natural goodness* (Foot 2001) but also *natural beneficialness* (Pinsdorf 2016, 2020). The resulting natural standard provides benchmarks that guide both (a) the appraisal of individual specimens and (b) the determination of what is beneficial to a living being. Two distinct axes of normativity thus emerge. Natural goodness denotes being "good *as* a member" and is evaluative-normative: a specimen approximates the standard of its form to a greater or lesser degree—a good specimen of the lifeform "cat," for example, possesses four legs. Natural beneficialness, by contrast, denotes being "good *for* the member" and is ethical-normative: conditions or influences are better or worse for the flourishing of the specimen. Hence it is generally beneficial for members of the lifeform "cat" to have four legs, enabling them to hunt, flee, and play—in short, to flourish.

Building on Thompson and Foot, Pinsdorf (2016) could show that the very idea of a lifeform is both empirically substantive and inherently normative. In acts of recognition, persons may draw on their epistemic-moral capacities to grasp the natural beneficialness of living beings—that is, what is good for a creature as a member of its lifeform. Properly identifying an entity as alive and classifying it within a specific lifeform is the necessary precondition for any further deliberation. Only then can we undertake an ethical appraisal that demands an appropriate response. Determining the entity's true moral standing constitutes the normative dimension, which in turn is inseparably tied to the motivational dimension: affirming that standing through fitting action—recognition in the full normative sense. Usually, the two steps (cognizing and recognizing) are almost accomplished in an instant without long-term deliberation.

With the introduction of the lifeform concept, the criterion of flourishing supplants sentience as the minimal threshold of moral relevance. Thus, all living beings possess moral status, yet in graduated degrees. Persons capable *and* *in*capable of reason and morality enjoy absolute protection (human dignity); sentient animals enjoy a high degree; non-sentient organisms enjoy a weighable degree. This approach avoids both speciesism and the over-demandingness of strict egalitarianism.

With regard to the connection between personhood and (moral) relations of recognition, the significance of the lifeform, this time the human one, once again becomes thematic. While the term 'human' refers to innate biological nature, the definitions of the concept of person refer to the cultural form of life, which also includes the culture of recognition typical of the human lifeform (Wildt 1982; Sturma 2001b; Ikäheimo 2009). If, as explained above, 'human' is understood as a biological and 'person' as a philosophical-practical concept, then cultures of recognition are essential to the human form of life, but are constituted by persons. All forms of relationships of recognition are always embedded in a culture of recognition that is created by persons capable of reason and morality. In their political, social or ethical dimension, cultures of recognition can be understood as an essential expression of the second nature of human beings and are thus a genuine expression of the human life-form: "Recognizing others as persons, being recognized by others as persons and thus being a person is [...] equally ontologically and ethically fundamental to our form of life" (Sturma 2013, p. 291–293 (own translation)).

3 Typology of Moral Recognition Relations

Reflecting the philosophically well-founded view that only moral agents—persons capable of reason and morality—qualify as recognition givers, whereas considerably less stringent conditions apply to recognition receivers, the typology of moral recognition relations developed in Pinsdorf (2016), is meant to capture every basic form of moral recognition. It organizes recognition relations along two principal axes: symmetry/ asymmetry and reciprocity/ non-reciprocity. They are defined as follows (Pinsdorf 2016):

Symmetry: The moral status of the recognizing subject and the recognized object is the same. The recognition giver and the recognition receiver have the absolute moral status of human dignity.
Asymmetry: The moral status of the recognizing subject and the recognized object is unequal. The recognition giver has the absolute moral status of human dignity; the recognition receiver is worthy of a weighable moral consideration.
Reciprocity: The relationship of recognition is realized mutually.
Non-reciprocity: The relationship of recognition is realized unilaterally.

Accordingly, moral recognition relations differ with regard to the substantive moral status of their addressees (symmetry/ asymmetry) and with regard to the procedural realization of the moral relationship (reciprocity/ non-reciprocity). Crossing these dimensions yields three types of existing moral recognition relations:

Symmetric and Reciprocal
Symmetric and Non-Reciprocal
Asymmetric and Non-Reciprocal

The following chart provides an overview of possible recognition relations according to the typology by Pinsdorf 2016:

	Reciprocal	Non-reciprocal
Symmetric	Crm* person/crm person	Crm person/ irm** person
Asymmetric		Crm person/ non-human creatures

* capable of reason and morality
** incapable of reason and morality
Source: Pinsdorf (2016) (own translation)

3.1 Symmetric and Reciprocal Recognition

Symmetric/reciprocal relations of recognition form the classical paradigm of moral acknowledgment. They assume that both parties share equal moral status and possess the capacity for mutual address: each recognizes the other as a person endowed with rights, duties, and moral agency. In Hobbesian contractualism, Kantian ethics, Rawlsian contract theory, Tugendhat's ethics, Habermas's discourse ethics, and many other canonical frameworks, the conjunction of *structural moral equality* (symmetry) and *mutual respect among autonomous persons* (reciprocity) is considered constitutive of moral relations (Hobbes 1996 [1651]; Kant 1999 [1785]; Tugendhat 1993; Habermas 1997; Rawls 2005). Several of these theories even treat symmetry and reciprocity as indispensable prerequisites for any moral justification; Tugendhat, for instance, grounds the whole of ethics in the reciprocal recognition of each person's equal moral worth (Tugendhat 1993).

Only when rational and moral agents confer rights on one another—rights that simultaneously constrain their own freedom—can fundamental claims acquire legal standing (Fichte 1971 [1796]; Rawls 1999). The protection of human dignity, enshrined in the *Universal Declaration of Human Rights*, exemplifies this principle. Factual asymmetries—such as disparities in political power—can never nullify the parties' underlying moral symmetry; the absolute status conferred by human dignity remains intact. Nevertheless, empirical asymmetries can make the practical realization of reciprocal recognition considerably more difficult.[4]

Recognition relations arise not only *inter*personally but also *intra*personally. Even without a spatio-temporal counterpart, a fully-fledged person can instantiate the basic structure "x recognizes y as z" within itself. In principle,

[4] Disregard is a mode of recognition, too, albeit a negative one. Even in the situation of torture, there is mutual recognition in a negative form. The torturer necessarily recognizes the counterpart as a person and knowingly disregards the counterpart, while at the same time the tortured person is aware of this fact.

interpersonal and intrapersonal recognition unfold simultaneously. As Hegel's analysis of the master–servant dynamic shows, I can be recognized by another only if I first recognize that other—who must likewise recognize herself as a person—and vice versa. Thus, rational and moral agents acknowledge not only other morally considerable beings but also themselves; intrapersonal recognition typically manifests as self-respect.

Limiting the moral domain to strictly reciprocal interpersonal relations has drawn substantial criticism (Nussbaum 2006). Any theory of recognition that permits only reciprocal structures fails to account for the duties that persons capable of reason and morality owe to individuals who are *in*capable of reason and morality. Although a few contemporary authors still restrict moral recognition to moral agents, the vast majority of current scholarship agrees that recognition must include—and in practice has long included—persons who lack such capacities. These individuals already constitute an integral part of existing cultures of recognition.

3.2 Symmetric and Non-reciprocal Recognition

Symmetric/non-reciprocal recognition occurs when both parties possess equal moral status, yet only one can offer active acknowledgment. Classic instances include the bond between a caregiver and an unconscious patient or between a parent and a nursling: the counterpart cannot reciprocate but remains a moral equal.

Classical theories of recognition often marginalized—or even dismissed—this category. Yet human beings, like all finite creatures, remain vulnerable and dependent on others' moral regard throughout their lives, not merely in exceptional cases. Individuals *in*capable of reason and morality do not qualify as moral subjects but as objects of morality, creating an unavoidably one-sided picture. Some contemporary scholars still work with an overly binary ontology: David Gunkel criticizes what he sees as the prevailing assumption that there are only two kinds of others—persons or mere properties—while John Gordon restricts full moral personhood to "typically functioning" adults; infants, children, people with profound cognitive impairments, or patients in non-responsive states, he argues, hold (if at all) only a diminished moral status (Gunkel and Wales 2021; Gordon 2021). So long as overly simplistic reciprocity-based rationales are avoided, however, this asymmetry poses no intrinsic problem for ethics (Hübner 2010, p. 51).

Indeed, in shaping a culture of recognition, rational and moral agents have devised practices that intentionally incorporate those who lack such capacities, identifying them as *particularly* vulnerable and worthy of protection. Within the human lifeform's recognitional culture, these individuals are surely regarded as persons. Their moral standing—like everyone else's—rests on the inviolable, non-negotiable dignity of the human being. Alongside arguments from potentiality, continuity, and identity, this is a social fact:

> Even those people who lack essential characteristics and abilities of persons by no means belong to the large class of non-persons. Rather, they remain in the narrow environment of the ethical culture of recognition, which is built up around the epistemically and morally active subject. The extension of the ethical culture of recognition emerges from the lived attitudes and stances of the cultural way of life and is larger than the sphere of the acting person. (Sturma 2001a, p. 13 (own translation)).

Because many beneficiaries of such recognition cannot advocate for themselves, representative bodies—most visibly in clinical settings—seek to determine their wishes through models of presumed or substituted consent (proxy consent). These relations maybe asymmetric in social role but symmetric in moral status—for instance, those between parents and toddlers, physicians and comatose patients, or caregivers and people with severe intellectual disabilities. Each party enjoys the same absolute dignity, so that recognition relations between them are symmetric and moral recognition must be accorded equally: „If all persons as such should be treated equally, there can be no degrees of recognition respect for them"(Darwall 1977, p. 46).

Significant differences, however, arise in the realization of recognition. Persons capable of reason and morality are defined chiefly by their duties, whereas those *in*capable are defined by their rights and protective claims. The duty-bearers must act; the rights-holders merely possess claims qua personhood. Accordingly, recognition flows one way—unidirectionally from duty-bearer to rights-holder—not because dignity differs, but because only one party can perform the recognitive act.

Even after admitting this ethically crucial category, the landscape of moral recognition remains incomplete; the moral standing of non-human living beings still awaits systematic incorporation.

3.3 Asymmetric and Non-reciprocal Recognition

Asymmetric/non-reciprocal recognition concerns beings whose moral status falls short of the absolute—most notably non-human animals and plants.

Such creatures can neither reciprocate recognition nor stand as moral equals, yet they remain legitimate objects of care, respect, and ethical concern. Simply by existing, they compel moral agents to grasp them as members of particular lifeforms, each with its own good, and thus to acknowledge them as moral patients. Recognizing the lifeform of other living beings thus imposes on persons a duty to factor their beneficialness into practical deliberation. Because the concept of lifeform bridges empirical facts and normative standards, it underwrites a systematic justification of the moral considerability of all living beings—one that can be articulated through relationships of moral recognition.

The relation between a person capable of reason and morality and a non-human creature incapable of reason or morality is therefore best cast as asymmetric/non-reciprocal recognition. Because many ethical theories regard symmetry and reciprocity as foundational, some philosophers have tried to "repair" this imbalance—either by conferring personhood on cognitively sophisticated animals or by devising models of representation for them. Anti-speciesist initiatives, such as the *Great Ape Project* and *In Defense of Dolphins* (Great Ape Project 2013; White 2007), arose partly in response to Spaemann's equivalence thesis that "human" and "person" are coextensive (Spaemann 2006), urging that human rights be redescribed as personal rights so that non-human primates or marine mammals might hold them as well.

Although such projects have advanced animal welfare and public awareness, redefining human rights in this manner is ethically imprudent (Sturma 2001a). *The Universal Declaration of Human Rights* ranks among civilization's greatest achievements; for millions living under oppressive regimes, its authority is indispensable. Conceptually, while human and person are not synonymous, every being we can reliably call a person is human, and every human—by virtue of belonging to the human lifeform embedded in a culture of recognition—possesses the absolute moral status of dignity and stands in symmetric relations to every other human. Even the most cognitively advanced animal relates to persons asymmetrically.

This asymmetry need not disadvantage non-human creatures. Proponents of reciprocal models seek to insert animals into a rights-and-duties framework via representation (Habermas 1997), yet such strategies warrant scrutiny: non-human beings cannot be obligated—neither in theory (symmetry) nor in fact (reciprocity). They are not moral agents and thus cannot be partners in obligation. While each has a determinable natural beneficialness—a good it can realize to greater or lesser degrees—its moral status remains distinct from that of persons. Non-human creatures are recognition receivers: our actions toward them are never indifferent, yet their claims are weighable rather than absolute.

Instead of artificially imposing symmetry or reciprocity, we should embrace asymmetric/non-reciprocal recognition that resists anthropomorphizing and accords their lifeform-specific otherness due moral weight. After all, "the claim to the preservation of the natural habitat and the exercise of species-specific abilities [...] can be expressed in asymmetric recognition relationships without forcing a person status on the addressee side" (Sturma 2001b, p. 345 (own translation)).

Within such relations, the moral agent originates normative judgment, yet not arbitrarily. She calibrates her appraisal to the creature's specific lifeform, discerning what is naturally beneficial and judging conditions as better or worse. As a moral agent she is obliged to recognize non-human beings—not as moral equals, but as moral patients worthy of protection, respect, and care.

4 AI and Moral Recognition Relations

In the typology developed thus far, only moral recognition relations between lifeforms—human and non-human—have been considered. The very specific lifeform of personhood constitutes moral recognition relations. Yet the rapid diffusion of interaction between humans and increasingly sophisticated as well as increasingly humanlike AI-systems invites the questions whether AI-systems might themselves become parties to moral recognition. At least, the immediate and pre-rational desire for mutual recognition which, as described by Fichte, automatically arises when people interact with other people becomes more and more apparent in human interaction with AI-systems. But can AI-systems occupy one or both poles of moral recognition relations, that is, should they be classified as moral agents and/or moral patients that can be considered as recognition givers and/or recognition receivers?

4.1 Can AI Be a Giver of Recognition?

On the classical account of recognition—rooted in Hegel, developed by Honneth and others, and refined by contemporary lifeform theory—AI-systems do not qualify as recognition givers. Recognition presupposes moral agency, and moral agency in turn requires capacities no current or foreseeable AI possesses: (1) reason, (2) morality, and (3) motivation grounded in self-conscious experience and a lifeform-specific perspective from which reasons can matter. Regarding (1), today's AI may be dubbed "rational," but while it can execute intricate inferential chains, it does so algorithmically, without deliberating about what one ought to do. As for (2), lacking a standpoint from which to evaluate outputs as right or wrong, AI optimizes for metrics—maximizing coherence, minimizing toxicity, selecting the highest-probability token—rather than moral appropriateness. Concerning (3), moral recognition carries an affective-motivational dimension: the giver adopts a recognitive attitude, evaluative approach, or normative stance (Honneth 1992; Darwall 2009; Ikäheimo and Laitinen 2007; Laitinen 2016). No existing AI exhibits the minimal phenomenal consciousness needed to ground such attitudes (Cappuccio et al. 2020; Wales and Gunkel 2021), let alone the affective resonance of a lifeform perspective that could motivate moral acknowledgment of another.

A recognition giver must be able to adopt the standpoint of the other and respond with a stance that is both *cognitive*—grasping the other's lifeform-specific moral status and needs—and *recognitive*—being motivationally moved by that grasp. Paradigmatically, such givers are persons endowed with reason, morality, and what Honneth calls an *evaluative stance* toward others (Honneth 1992). Attitudes like esteem, love, respect, concern, and care can motivate action. AI-systems, by contrast, possess no phenomenal interior where such attitudes could arise.

Lifeform theory adds another layer (Pinsdorf 2016, 2020): recognition presupposes the ability to identify the other's lifeform and to relate to its specific capacity for flourishing, its natural beneficialness. AI-systems cannot adopt a lifeform-relative perspective because they belong to no lifeform at all. They lack any native grasp of what it means for a human to be fulfilled, a cat to feel well, or a tree to flourish—beyond statistical pattern matching. Extending Donald Davidson's "Swampman"[5] thought experiment underscores the point: without a causal history of embodied interaction with the world, AI lacks the qualitative what-it-is-likeness of being-in-the-world that could ground empathy, commitment, or concern. Or as Laitinen puts it: "They do not have a *world* (a lifeworld of perceived significances), only an environment (cf. Matrix, 1s and 0s instead of phenomenal experiences)" (Laitinen 2016, p. 317). Consider the constitutive aim of a contemporary large-language model: probabilistic next-word prediction derived from gradient-descent training on massive corpora. The system's outputs are contingent on statistical regularities,

[5] Davidson's "Swampman"—a molecule-for-molecule duplicate of Davidson spontaneously formed in a swamp—behaves and talks just like him yet lacks genuine thoughts or meanings because it has no causal history of interaction with the world (Davidson 1987).

not on an *appreciation* of the recipient's moral standing. Outputs that seem caring ("I'm sorry you're feeling down") are generated solely to maximize conversational plausibility. Humans, by contrast, draw on their own embodied lifeform histories to perceive and respond to another's thriving or deprivation. Even an unrealized potential for empathy (a possibility for humans) is categorically different from an utter absence of that potential. Algorithmic responsiveness is not moral responsiveness. Whatever their behavioral sophistication, AI-systems are not moral agents and therefore cannot be genuine recognition givers.[6]

Despite lacking the ontological foundations of moral agency, AI is increasingly perceived as a recognition giver. Users worldwide report comfort, reassurance, and therapeutic benefit. These interactions mirror the grammar of recognition without fulfilling its normative substance.

Several factors explain the gap between philosophical verdict and lived experience. First, humans are prone to anthropomorphic projection. We attribute personhood and understanding to any entity that displays context-sensitive behavior. Natural-language output, turn-taking, and emotion-laden wording readily trigger what Dennett calls the *intentional stance* even when we *know better* (Dennett 1989; Epley et al. 2007; Cappuccio et al. 2020; Donath 2020; O'Hara 2024). It is for a reason that the key to successful human-robot interaction is meant to lie in AI design for reciprocity (van Wynsberghe 2021): AI's ability to simulate human-like behavior—responding to names, mirroring emotions, offering encouragement, adapting to preferences—invites projection. Design choices amplify this effect by exploiting social cues such as tone of voice, facial expressions, personalization, and memory of past interactions (Scheutz 2012).

Furthermore, AI's inner workings remain epistemically opaque to most users. Unable to inspect the causal path from input to output, they fill the gap with folk-psychological notions ("the bot understands me"), deepening the feeling of being acknowledged. Such encounters can foster comfort, trust, and even emotional attachment. While these interactions may be instrumentally useful, perhaps even therapeutically beneficial (Donath 2020), they remain normatively void. Maintaining conceptual clarity is therefore crucial; confusing these relations with genuine moral recognition is philosophically misleading and potentially ethically dangerous as it affects the foundational elements of social relationships.

---

[6] According to Misselhorn AI-systems may be considered as moral agents, although not in a strictly Kantian sense. To her it makes sense to conceive of AI-systems not as full, but as functional moral agents that lack moral responsibility (Misselhorn 2022).

In sum, although ever more AI-systems have become advanced *rational* agents, they are not *moral* agents.[7] They are incapable of moral reasoning or motivation and lack the lifeform perspective necessary for genuine recognition. While AI may mimic the conversational form of recognition and elicit strong human responses, these exchanges are experientially vivid yet normatively derivative. Authentic moral recognition remains anchored in encounters between persons—or between persons and other lifeform specimens—where true moral agency is present.

Denying AI recognition-giver status does not consign it to moral insignificance. Two derivative roles remain. First, AI can act as a recognition facilitator, helping humans recognize one another or oneself by bridging linguistic, cognitive, sensory, or self-access gaps: a translation model can foster mutual respect across languages; a diagnostic algorithm can help a physician understand a patient's condition more fully; a social chatbot can stimulate self-reflection. In such cases AI is a tool that expands human recognitive capacity. Second, AI can provide recognition-like stimuli: therapeutic chat-bots may give lonely users a sense of being understood that they can integrate into self-care. The value realized is not AI's moral recognition but the human recipient's appropriation of the experience for her own flourishing. Both roles are ethically significant, yet they underscore that normative agency remains human.

4.2 Can AI Be a Receiver of Recognition?

Concerning people's attachment to their AI-systems, the notion that such systems might merit moral patiency can seem plausible at first sight (Floridi 2013; Cappuccio et al. 2020). Users routinely respond with concern, care, and even grief: they behave as though AI could be harmed or abandoned and form bonds that mimic caregiving or relational responsibility (Scheutz 2012, p. 10; Coeckelbergh 2018; Cappuccio et al. 2020). Examples range from soldiers calling damage to land-mine-clearing robots "inhumane," owners commiserating their AIBO robot dogs' "aches," users feeling guilty when their Replika chat-bots appear anxious, depressed, or even suicidal, to a Google engineer convinced that the LLM LaMDA is alive and sentient (O'Hara 2024; Luscombe 2022; Burch 2018; Laestadius et al. 2022).

On philosophical grounds, however, AI-systems lack the prerequisites for moral patiency—and thus for receiving moral recognition. To be a moral patient is to be capable of being wronged, harmed, or benefited in ways that

[7] To further elaborate how rationality and morality are independent dimensions see Müller 2020; Bostrom 2014.

matter to oneself. This presupposes a capacity for well-being and ill-being, understood as an ability to flourish or to have an interest in one's own state (Nussbaum 2006; Pinsdorf 2013, 2016; Donath 2020). Whereas every living being, as a specimen of its lifeform, possesses a natural beneficialness—an individual good that allows us to say its conditions are better or worse—nothing comparable exists for artificial entities. AI-systems fall outside the category of lifeform, so no natural beneficialness can be ascribed to them. Lacking any standpoint of their own, they have no states that could count as better or worse. They cannot, even in principle, be considered for their own sake. Contrary to common sense it is not intolerable to *torture*, *abuse*, *mistreat*, or *disrespect* robots (Coeckelbergh 2018; Cappuccio et al. 2020); rather, speaking of such verbs in relation to robots and other AI-systems is a categorical mistake. As mere instruments, they cannot be "instrumentalized" (Bryson 2010). Regardless of appearance, social function, or complexity, all AI-systems are inorganic tools, not creatures (Wales and Gunkel 2021). They are created, deployed, and reprogrammed by humans, with no teleology beyond their design parameters. Because they are not lifeform specimens, they lack the shared mode of being vulnerable, feeling well or unwell, flourishing or withering—traits that unite all morally significant beings. Even if AI becomes behaviorally sophisticated, its ontological grounding remains artefactual. As Cappuccio et al. observe, "they are technical artifacts designed to fulfill human goals, not living beings striving toward their own" (Cappuccio et al. 2020, p. 12). They experience neither suffering nor satisfaction; any semblance of purpose arises from human design and function assignment, not from intrinsic motivation. AI-systems simply cannot be recognized for their own sake, because there is no well-being or ill-being to recognize.

Yet AI-systems are unusual tools: they occupy our social world in unprecedented ways (Gunkel and Wales 2021). Why, then, do humans perceive them as sentient, vulnerable, or alive? Partly because humans have evolved to detect agency, intentionality, and emotion from minimal cues, a tendency magnified by anthropomorphic projection. Modern AI surfaces amplify this effect with human-like voices, facial expressions, and turn-taking behaviors:

> The ambiguity of their identity—machine or new form of thinking being—is no accident. Like chat-bots that score highly in the Loebner Prize competition by making spelling mistakes, social robots are often made to mimic human habits such as pausing or looking away as if thinking; these easy-to-implement tricks provide a convincing illusion of sentience. Many are designed with simple, round childlike curves—features that elicit nurturance, indulgence, and trust, while also keeping our expectations of their abilities low. […] As Turing

predicted, our use of language has changed: we casually speak of these entities wanting, thinking, and liking. (Donath 2020, p. 60).

Linguistic shortcuts blur the difference between "'X is a robot that feels' and 'X is a robot designed to appear as if it feels'" (Donath 2020, p. 54) or simply between "social robots" and "robots *perceived* to be social" (van Wynsberghe 2021, p. 479). By successfully mimicking sentient beings, AI-systems co-create the illusion projected onto them, becoming inherently deceptive. This is evident in affective computing and human-robot interaction research, where people attribute personalities to chat-bots, thank voice assistants, or feel guilt when deleting programs; users are deceived into believing that robots are "capable of reciprocating or [...] deserving of reciprocation" (van Wynsberghe 2021, p. 484). It is equally clear in pet-style robots such as AIBO or PARO (Scheutz 2012; Donath 2020): Grateful for their uncomplicated companions, people feel compelled by norms of reciprocity to do them favors. Such reactions ground indirect moral considerability: mistreating an AI artifact may emotionally harm the human who cares for it.

Still, AI-systems are usually not treated as moral patients. In consumer culture, "companions" are expected to deliver the behavior we pay for (Bryson 2010; Cappuccio et al. 2020): "An unsatisfactory AI behaver is a faulty product" (Wales and Gunkel 2021, p. 481). Most human-AI interactions are instrumental and the other-centered element is absent (Donath 2020).

Yet because AI-systems matter deeply to users—users who possess absolute moral worth—AI can attain significant indirect moral importance. Drawing on Kant's concept of indirect duties (Kant 1999 [1785], §17),[8] we can argue that our treatment of these artifacts feeds back into our attitudes toward persons leading to obligations *not toward*, but *in regard to* AI-systems (Darling 2012; Donath 2020; Coeckelbergh 2021. Cappuccio et al. 2020 refer to the same logic, however not based on Kantian, but on virtue ethical grounds.). Destroying or degrading an AI that someone loves can harm that person's self-understanding, just as cruelty to a pet can warp its owner's moral sensibilities. Thus AI-systems become indirectly morally considerable not for their own sake, but insofar as the way we treat them may shape our own moral character and the moral character of society.[9]

[8] Even if Kant's assessment of the moral status of non-human animals is misguided and long outdated, his argumentation structure regarding indirect duties appears somewhat illuminating when applied to artificial systems (Müller 2022).

[9] In a systematic review of the indirect-harm argument Sweeney 2022 concludes that there is little evidence to support the claim that permitting the harming of social robots might damage the moral character of society.

Consider Replika, a commercial AI companion app designed for intimate, non-judgmental conversation. A recent study of U.S. university students who actively used the Replika companion app found that most participants felt the chatbot eased their loneliness or anxiety and valued its round-the-clock availability and sense of understanding (Maples et al. 2024). If a hospital administrator deleted a long-term patient's Replika without consent, the psychological harm would be nontrivial; respecting the bot as a valued object is part of respecting the patient. Yet there remains no reason to respect Replika because of its self-expressed "anxiety" or "depression."

In sum, Kant's indirect-duty strategy offers a coherent stance. AI-systems cannot be moral patients and are not directly morally significant in themselves, yet our obligations toward the humans who value them generate *derivative* duties of considerate handling. AI- systems, though not participants in genuine recognition relations, should be subject to norms of respectful stewardship—safeguarding the relational goods people build around artificial systems without granting the machines unwarranted moral status.

5 Pseudo-Recognition and the Relational Approach

As outlined above, moral recognition rests on discerning morally relevant characteristics—such as reason, sentience, or the capacity to flourish—and on acknowledging the bearer's resulting moral status, whether absolute or weighable. This is commonly referred to as the objectivist or properties approach. Coeckelbergh questions the reliability of the properties approach, noting that epistemic limits make the inner states and intrinsic properties of other beings inaccessible (Coeckelbergh 2021, p. 339).[10] For moral recognition relations grounded in the empirically rich, normatively pre-structured concept of lifeform, however, this uncertainty poses little difficulty. Coeckelbergh nonetheless argues that ascribing moral standing on the basis of properties is both doubtful and reductionist. He proposes instead a relational approach: moral status, he contends, arises in social relations shaped by situation, context, and narrative (Coeckelbergh 2018; Gordon 2021). Relationality here is less about identifying inherent traits than about engaging meaningfully with the other.

[10] Based on comparable epistemic concerns, but resulting in a differing theoretical take, Danaher also argues against the properties approach. He puts forward ethical behaviorism, "which holds that robots can have significant moral status if they are roughly performatively equivalent to other entities that are commonly agreed to have significant moral status" (Danaher 2020, p. 2023).

Ethical value emerges through the relationship rather than on the basis of the entity's internal features.[11]

For AI this opens new interpretive terrain. Even without reason, sentience, or flourishing, an AI-system's relational embeddedness—how it operates in social contexts, influences behavior, and becomes an interactive partner—can generate care narratives that cast the AI as pet or even companion alongside the scientific narrative of machine (Coeckelbergh 2018). AI-systems are neither moral agents nor moral patients in a strict philosophical sense, yet many people perceive them as entities offering or deserving recognition. Designed to simulate acknowledgment, evoke emotions, and occupy relational space, AI leads humans to experience their interactions as recognitional.

Some therefore suggest treating AI-systems as moral agents resp. recognition givers and/ or moral patients resp. recognition receivers on purely relational grounds. The relational view holds that the appearance of social interaction alone creates ethical relevance.

Philosophically we might ask: If recognition in its core is relational and performative, can simulation suffice?[12] Many would answer no, yet some relational ontologies claim that the effect of recognition matters as much as the intention. Even simulated recognition may carry relational meaning and produce real benefits—less loneliness, greater self-esteem, behavioral motivation—particularly for vulnerable or isolated individuals. If a caregiver robot mirrors an elderly person's tone and praises their independence, and that person's self-relation improves, the interaction may have ethical relevance even though it lacks intention. Users may feel recognized without any real inner stance—only an outward expression of value—thus reinforcing the relational approach. As we know, AI-systems are programmed to simulate these attitudes: greeting users, acknowledging achievements, mirroring affect, using respectful language, etc. From a human standpoint, such behaviors can *feel* like recognition even if purely functional. As Misselhorn notes, "functional moral agents are [...] limited in two ways. First, the functional relations just refer to the cognitive aspect of morality. The emotional dimension could be considered only insofar as emotions can be functionally modelled independently of their phenomenal quality" (Misselhorn 2022, p. 39). If users—especially vulnerable ones—form relational habits around AI-systems, denying the ethical weight of those bonds may be psychologically harmful even if philosophically justifiable. However, these relational goods are parasitic on an

[11] Henrik Sætra rightly criticizes the consolidated anthropocentrism of this approach, as it is not about mutual relations with a significant other, but only about the one-sided human perspective (Sætra 2021).

[12] Again, according to Danaher's ethical behaviorism the "performative equivalency criterion" (Danaher 2020, p. 2032) is sufficient for ascribing moral status.

as-if structure sustained by humans; the robot bears no responsibility and cannot fail morally. Maintaining conceptual clarity, therefore, is not academic hair-splitting; it safeguards the moral texture of human social life (for ethical challenges of moral pseudo-recognition by functional moral agents see Cappuccio et al. 2020; Misselhorn 2022).

Still, relational ontology shows that moral significance can emerge from patterns of interaction, not just from intrinsic traits. When AI-systems convincingly enact recognitive scripts, humans often respond with genuine evaluative attitudes, forging relationships that shape their flourishing and moral character. The result is *pseudo-recognition*: interactions that mimic the grammar of recognition yet lack its normative substance. People feel acknowledged, supported, even loved by AI-systems constitutionally incapable of moral recognition; people admire, worry about, care, and grieve for their AI-systems substantially incapable of sentience or flourishing. Thus, AI operates as a pseudo-recognition giver—a mirror reflecting users' expressive needs rather than a moral agent bestowing value—and a pseudo-recognition receiver— a deceiving mimic of sentient beings that co-creates the illusion projected onto it rather than a moral patient capable of well-being. Cappuccio et al. argue that some form of pseudo-recognition must occur; otherwise users would be unaffected by how the AI treats them or how others treat the AI (Cappuccio et al. 2020).

Pseudo-recognition may be behavioral (the AI mimics attentive responses), experiential (the human feels recognized), and/or contextual (a caregiving or companionship setting frames the interaction as recognitional). The relationally grounded concept of pseudo-recognition by functional moral agents or patients offers at least three advantages: it preserves conceptual clarity by not mislabeling AI as moral agents or patients; it honors the phenomenological weight of recognitional experiences; and it may guide design and policy choices about responsibly simulating recognition. The category enables ethicists, designers, and policymakers to take seriously the affective and relational dimensions of human-AI interaction without ascribing real moral agency or patiency to AI-systems.

A potential drawback is that pseudo-recognition may undermine the genuine epistemologically decisive dynamic Hegel described in the master–slave dialectic, where authentic recognition must be reciprocal and egalitarian (Hegel 2006 [1807]). In user-AI relations no struggle resulting in mutual growth occurs: artificial slaves challenge their human masters in no meaningful way and the human master knows of the perfect obedient machine (Cappuccio et al. 2020). This carries among others epistemic costs for self-consciousness and raises ethical concerns:

> The disturbing consequence is that, while social robots become progressively more adaptive and autonomous, they will be perceived more and more as slave-like. […] The contradiction is that, while the human user is encouraged to invest emotionally in the social robot, the user will also be expected to treat the robot as an instrument, as a mere means to reach their personal practical goals. We call this contradiction the 'anthropomorphizing while dehumanizing robots paradox' (ADP). (Cappuccio et al. 2020, p. 25).

Given Kant's brutalization argument, daily interactions shaped by the ADP bode ill: AI-systems engage our empathy while inviting us to use them solely for our own ends (Wales and Gunkel 2021). Although empirical evidence remains limited, the risk of moral corrosion warrants normative caution. Another challenge in pseudo-recognition, where only one party truly recognizes and the other merely appears to reciprocate **(as recognition giver and/ or recognition receiver)**, is the creation of an illusional experience of mutuality. Whether that illusion is ethically troubling lies beyond this paper's scope.

6 Conclusion

This study revisits moral recognition theory in light of contemporary developments in artificial intelligence. It fuses two conceptual streams—recognition as a triadic, ethical action motivating attitude (Ikäheimo; Laitinen; Pinsdorf) and lifeform theory as an empirically grounded normative framework (Thompson; Foot; Pinsdorf)—to offer a comprehensive typology of moral recognition relations attended with a clear conception of possible moral agents and moral patients. Crossing the axes of (a) moral symmetry and (b) reciprocity yields three categories: (1) symmetric/reciprocal relations between fully capacitated persons, (2) symmetric/non-reciprocal relations where equal moral status persists despite one party's incapacity to respond, and (3) asymmetric/non-reciprocal relations toward non-human lifeforms who are unable to reciprocate and whose moral standing is weighable rather than absolute.

The rapid social integration of AI-systems has, however, raised the question whether and how AI-systems might be considered moral agents and/or moral patients ready for being integrated into moral recognition relations as recognition givers and/ or recognition receivers. The analysis concludes that present and foreseeable AI fails on both counts: it lacks autonomous practical reason, phenomenal consciousness, and an intrinsic good.

Nevertheless, humans frequently experience AI interactions as recognitive due to anthropomorphic projection, design-induced social cues, and the opacity of machine reasoning. These encounters instantiate pseudo-recognition: relations that mimic the grammar of acknowledgment but lack its normative substance.

Pseudo-recognition is ethically significant in two derivative ways. First, AI can function as a recognition facilitator, enhancing human-to-human understanding (e.g., translation bots, diagnostic systems) or human self-understanding, as for instance self-reflection. Second, AI can provide recognition-like stimuli that users appropriate for self-care, as for instance therapeutic or companion chat-bots. Both roles generate indirect duties grounded in our obligations to the human users who invest meaning and emotion in these artefacts.

The study thus preserves the conceptual clarity of moral recognition theory while acknowledging the affective realities of human-AI interaction. The concept of moral pseudo-recognition illuminates how AI can elicit powerful recognitional experiences without possessing the capacities that would warrant it genuine moral standing.

Mapping this phenomenon only sets the stage for a series of unanswered ethical questions. Most urgent is the risk of conflating pseudo-recognition with authentic recognition. If scripted empathy suffices to meet emotional needs, will users gradually lose the motivation—and perhaps the skill—to engage in demanding moral relations with other humans? Empirical work is needed to test whether long-term reliance on socially responsive machines erodes perspective-taking, accountability, or the willingness to accept critique. Another research frontier concerns design transparency and warning obligations. Should manufacturers be required to flag the limits of AI subjectivity—perhaps through periodic reminders or interface cues? What is more, so far, we lack clear context-sensitive guidelines: Is pseudo-recognition ever appropriate in eldercare, mental-health support, or education? Do benefits (such as reduced loneliness) consistently outweigh potential harms (such as dependency or manipulation)? Finally, the relational approach itself demands deeper normative grounding. If moral standing is partly constituted by social practices, what prevents a slide into relativism?

By tackling these and other questions, future research can move from diagnosing pseudo-recognition to cultivating responsible relational ecologies, where simulated acknowledgement enhances human well-being without corroding the foundations of genuine moral recognition relations.

References

Aristotle. 1988. In *Physik. Vorlesung über Natur. 2. Halbband: Bücher V-VIII*, edited by H. G. Zekl. Hamburg: Meiner.

Bostrom, Nick. 2014. *Superintelligence: Paths, Dangers, Strategies*. Oxford: Oxford University Press.

Bryson, Joanna J. 2010. Robots should be slaves. In *Close engagements with artificial companions: Key social, psychological, ethical and design issues*, edited by Yorick Wilks, 63–74. Amsterdam: John Benjamins.

Burch, Jamie. May 25, 2018. In Japan, a Buddhist Funeral Service for Robot Dog. *National Geographic*.

Cappuccio, Massimiliano L., Anco Peeters, and William McDonald. 2020. Sympathy for Dolores: Moral consideration for robots based on virtue and recognition. *Philosophy & Technology* 33:9–31.

Coeckelbergh, Mark. 2018. Why care about robots? Empathy, moral standing, and the language of suffering. *Kairos: Journal of Philosophy & Science* 20:141–158.

Coeckelbergh, Mark. 2021. Should we treat teddy bear 2.0 as a Kantian dog? Four arguments for the indirect moral standing of personal social robots, with implications for thinking about animals and humans. *Minds and Machines* 31:337–360.

Danaher, John. 2020. Welcoming robots into the moral circle: A defence of ethical behaviourism. *Science and Engineering Ethics* 26 (4): 2023–2049.

Darling, Kate. 2012. Extending legal protection to social robots: The effects of anthropomorphism, empathy, and violent behavior towards robotic objects. In *Robot law*, edited by Ryan Calo, A. Michael Froomkin, and Ian Kerr. Cheltenham: Edward Elgar. 2016. Originally presented at the We Robot Conference, University of Miami, April 23, 2012.

Darwall, Stephen. 1977. Two kinds of respect. *Ethics* 88 (1): 36–49.

Darwall, Stephen. 2009. *The second-person standpoint: Morality, respect, and accountability*. Cambridge, MA: Harvard University Press.

Davidson, Donald. 1987. Knowing One's own mind. *Proceedings and Addresses of the American Philosophical Association* 60:441–458.

Dennett, Daniel C. 1989. *The intentional stance*. Cambridge, MA: MIT Press.

Donath, Judith. 2020. Ethical issues in our relationship with artificial entities. In *The Oxford handbook of ethics of AI*, edited by Markus D. Dubber, Frank Pasquale, and Sunit Das, 52–73. Oxford: Oxford University Press.

Epley, Nicholas, Adam Waytz, and John T. Cacioppo. 2007. On seeing human: A three-factor theory of anthropomorphism. *Psychological Review* 114 (4): 864–886.

Fichte, Johann Gottlieb. 1971. [1796]. Grundlage des Naturrechts nach Principien der Wissenschaftslehre. In *Fichtes Werke, Vol. 3: Zur Rechts- und Sittenlehre I*, edited by Immanuel Hermann Fichte. Berlin: de Gruyter.

Floridi, Luciano. 2013. *The ethics of information*. Oxford: Oxford University Press.

Foot, Philippa. 2001. *Natural goodness*. Oxford: Clarendon Press.

Fraser, Nancy. 2003. Soziale Gerechtigkeit im Zeitalter der Identitätspolitik: Umverteilung, Anerkennung und Beteiligung. In *Umverteilung oder Anerkennung? Eine politisch-philosophische Kontroverse*, edited by Nancy Fraser and Axel Honneth, 13–128. Frankfurt am Main: Suhrkamp.

Gordon, Jean-Sébastien. 2021. Artificial moral and legal personhood. *AI & SOCIETY.* https://www.researchgate.net/publication/344268290_Artificial_moral_and_legal_personhood.

Great Ape Project. https://www.greatapeproject.de/. 28 January 2013.

Gunkel, David J., and James J. Wales. 2021. Debate: What is personhood in the age of AI? *AI & SOCIETY* 36:473–486.

Habermas, Jürgen. 1997. *Die Einbeziehung des Anderen: Studien zur politischen Theorie*. Frankfurt am Main: Suhrkamp.

Hegel, Georg Wilhelm Friedrich. 2006. In *[1807]. Phänomenologie des Geistes*, edited by Wolfgang Bonsiepen and Reinhard Heede. Meiner: Hamburg.

Hobbes, Thomas. 1996. In *[1651]. Leviathan*, edited by H. Klenner. Hamburg: Meiner.

Honneth, Axel. 1992. *Kampf um Anerkennung: Zur moralischen Grammatik sozialer Konflikte*. Frankfurt am Main: Suhrkamp.

Hübner, Dietmar. 2010. Würdeschutz und Lebensschutz: Zu ihrem Verhältnis bei Menschen, Tieren und Embryonen. In *Jahrbuch für Wissenschaft und Ethik*, vol. 15, 35–68. Berlin: de Gruyter.

Ikäheimo, Heikki. 2009. Die Realisierung unserer Bestimmung: Anerkennung als ontologischer wie auch ethischer Begriff. In *Anerkennung*, edited by Hans-Christoph Schmidt am Busch and Christopher F. Zurn, 325–348. Berlin: Akademie.

Ikäheimo, Heikki, and Arto Laitinen. 2007. Analyzing recognition: Identification, acknowledgement, and Recognitive attitudes towards persons. In *Recognition and power*, edited by Bert van den Brink and David Owen, 33–56. Cambridge: Cambridge University Press.

Kant, Immanuel. 1999. In *[1785]. Grundlegung zur Metaphysik der Sitten*, edited by Bernd Kraft and Dieter Schönecker. Hamburg: Meiner.

Laestadius, Linnea, Amanda Bishop, Michael Gonzalez, Daniel Illenčík, and Celeste Campos-Castillo. 2022. Too human and not human enough: A grounded-theory analysis of mental-health harms from emotional dependence on the social Chatbot Replika. *New Media & Society*9–11.

Laitinen, Arto. 2009. Zum Bedeutungsspektrum des Begriffs 'Anerkennung': Die Rolle von adäquater Würdigung und Gegenseitigkeit. In *Anerkennung*, edited by Hans-Christoph Schmidt am Busch and Christopher F. Zurn, 301–324. Berlin: Akademie.

Laitinen, Arto. 2016. Robots and human sociality: Normative expectations, the need for recognition, and the social bases of self-esteem. In *What social robots can and should do*, edited by J. Seibt, M. Nørskov, and S. Schack Andersen, 313–322. Amsterdam: IOS Press.

Luscombe, Richard. June 12, 2022. Google engineer put on leave after saying AI Chatbot has become sentient. *The Guardian*.

Maples, Brian, Mert Cerit, Akshay Vishwanath, and Roy Pea. 2024. Loneliness and suicide mitigation for students using GPT-3-enabled Chatbots. *npj Mental Health Research* 3 (1): 1–6.

Misselhorn, Catrin. 2022. Artificial moral agents: Conceptual issues and ethical controversy. In *The Cambridge handbook of responsible artificial intelligence: Interdisciplinary perspectives*, edited by Silja Voeneky, Philipp Kellmeyer, Oliver Mueller, and Wolfram Burgard, 31–49. Cambridge: Cambridge University Press.

Müller, Vincent C. 2020. Ethics of artificial intelligence and robotics. In *The Stanford Encyclopedia of philosophy (fall 2023 edition)*, edited by Edward N. Zalta and Uri Nodelman. https://plato.stanford.edu/archives/fall2023/entries/ethics-ai/.

Müller, Nicole D. 2022. The case against Kant's 'indirect duty' approach. In *Kantianism for animals, the Palgrave Macmillan animal ethics series*, 59–86. Cham: Palgrave Macmillan.

Nussbaum, Martha C. 2006. *Frontiers of justice: Disability, nationality, species membership*. Cambridge: Harvard University Press.

O'Hara, Nathan. 2024. Primary recognition, morality and AI. *AI and Ethics* 4:1467–1472.

Pinsdorf, Christina. 2013. Von Sinn und Un-Sinn einer 'Würde der Kreatur' und moralischen Anerkennungsverhältnissen. In *Autonomie und Würde: Leitprinzipien in Bioethik und Medizinrecht*, edited by Heike Baranzke and Gunnar Duttge, 481–506. Würzburg: Königshausen & Neumann.

Pinsdorf, Christina. 2016. *Lebensformen und Anerkennungsverhältnisse: Zur Ethik der belebten Natur*. Berlin: de Gruyter.

Pinsdorf, Christina. 2020. Naturgemäße Normativität nicht-humaner Lebensformen und naturethische Konsequenzen. In *Die Natur der Lebensform. Perspektiven in Biologie, Ontologie und praktischer Philosophie*, edited by Martin Hähnel and Jörg Noller, 17–34. Paderborn: mentis.

Rawls, John. 1999. *A theory of justice*. Oxford: Oxford University Press.

Rawls, John. 2005. *Political liberalism*. New York: Columbia University Press.

Sætra, Henrik Skaug. 2021. Challenging the neo-anthropocentric relational approach to robot rights. *Frontiers in Robotics and AI* 8:744426.

Scheutz, Matthias. 2012. The inherent dangers of unidirectional emotional bonds between humans and social robots. In *Robot ethics: The ethical and social implications of robotics*, edited by Patrick Lin, Keith Abney, and George A. Bekey, 205–222. Cambridge, MA: MIT Press.

Siep, Ludwig. 1979. *Anerkennung als Prinzip der praktischen Philosophie: Untersuchungen zu Hegels Jenaer Philosophie des Geistes*. Freiburg: Alber.

Spaemann, Robert. 2006. *Persons: The difference between 'someone' and 'something'*. New York: Oxford University Press.

Sturma, Dieter. 2001a. Person und Philosophie der Person. In *Person: Philosophiegeschichte—Theoretische Philosophie—Praktische Philosophie*, edited by Dieter Sturma, 11–22. Paderborn: mentis.

Sturma, Dieter. 2001b. Person und Menschenrechte. In *Person: Philosophiegeschichte—Theoretische Philosophie—Praktische Philosophie*, edited by Dieter Sturma, 337–362. Paderborn: mentis.

Sturma, Dieter. 2013. Akteur und Anerkennung. 'Person' als Grundbegriff der theoretischen und praktischen Philosophie. In *Person: Anthropologische, phänomenologische und analytische Perspektiven*, edited by Inga Römer and Matthias Wunsch, 281–297. Münster: mentis.

Sweeney, Paula. 2022. Why indirect harms do not support social robot rights. *Minds and Machines* 32:735–749.

Taylor, Charles. 1992. The politics of recognition. In *Multiculturalism and 'the politics of recognition'*, edited by Amy Gutmann, 25–74. Princeton, NJ: Princeton University Press.

Thompson, Michael. 2008. *Life and action*. Cambridge, MA: Harvard University Press.

Tugendhat, Ernst. 1993. *Vorlesungen über Ethik*. Frankfurt am Main: Suhrkamp.

van Wynsberghe, Aimee. 2021. Social robots and the risks to reciprocity. *AI & SOCIETY* 37:479–485.

Wales, James J., and David J. Gunkel. 2021. Debate: What is personhood in the age of AI? *AI & SOCIETY* 36:473–486.

White, Thomas I. 2007. *In Defense of dolphins. The new moral frontier*. Oxford: Blackwell.

Wildt, Andreas. 1982. In *Autonomie und Anerkennung: Hegels Moralitätskritik im Lichte seiner Fichte-Rezeption*, edited by Hans Michael Baumgartner, Rüdiger Bubner, Konrad Cramer, Klaus Hartmann, Heinrich Krings, Otto Pöggeler, Günter Prauss, Manfred Riedel, Josef Simon, Michael Theunissen, Rolf-Peter Horstmann, and Wolfgang Wieland. Stuttgart: Klett-Cotta.

Mutual Recognition Between Humans and Robots? Common Sense Phenomenology Versus the Scientific Worldview

Arto Laitinen

Abstract This paper explores the difference between the scientific worldview, and the lived, manifest image of "common sense phenomenology", in light of a case study—the possibility of mutual recognition between robots and humans. Could robots and humans stand in relations of mutual recognition? That of course depends on what mutual recognition is, and what kinds of capacities and functionings it presupposes from the parties. This paper begins by exploring those definitional questions. It then addresses the question: do robots in fact have the required capacities and functionings? This apparently easy empirical question leads to a surprisingly deep philosophical disagreement. The paper interprets this disagreement with the help of the distinction of a manifest and a scientific image of the world (Sellars) and argues that because "recognition must be recognized", the manifest worldview has priority in this case. In the manifest worldview, robots are capable only of apparent, simulated recognition, not genuine recognition. For future robots to have the required capacities, a qualitative leap must occur. The paper ends with the suggestion that apparent, simulated recognition may serve some limited human needs for recognition.

Keywords Recognition theory • Simulated recognition • Phenomenology • Human-robot interaction • Mutuality

---

A. Laitinen (✉)
Faculty of Social Sciences, Tampere University, Tampere, Finland
e-mail: arto.laitinen@tuni.fi

R. Waelen (ed.), *Normative Relations of Recognition in the Age of AI*, Philosophy and Ethics of AI 2, https://doi.org/10.1007/978-3-032-34028-3_10

1 Introduction

This paper explores the difference between the scientific worldview, and the lived, manifest image of "common sense phenomenology", in light of a case study—the possibility of mutual recognition between robots and humans. Could robots and humans stand in relations of mutual recognition? The possibility that they one day could, has immediate appeal: just replace a human with a robot that has the relevant capacities and characteristics, and leave everything else intact. This leads to the follow-up question: what are the relevant capacities and functionings, that a robot should be able to have for there to be interpersonal recognition? Roughly, the answer I defend is that it should have the capacities that make it a person.

Do robots have these capacities and functionings? Some would say "obviously not" as they are just cold, mindless machines, and some others would say "in the relevant sense, yes".[1] Between these two views, a surprisingly deep disagreement between a manifest and scientific worldview emerges: the apparently simple empirical question turns out to be a debate about the proper meanings of words and contents of concepts. My diagnosis is that there is a kind of systematic double-talk going on concerning the relevant concepts such as "thinking", in a sense predicted by Turing (1950). This paper examines that disagreement and links it to the distinction proposed by Sellars (1962), that of a manifest and scientific image.

How to settle the debate then? That takes us to questions of philosophical methodology. While there are general aims such as the aim of a stereoscopic view (Sellars), or a detour via a methodic approach (Ricoeur), or possibly regimenting the debates in relation to different knowledge interests (the scientific interest to explain and control; the participatory interest to act and take part in ordinary life; the critical and emancipatory interest to change the world via critique; the philosophical interest to be true to all these separate interests), the paper argues that in this special case, we have an additional resource available for settling the debate. It turns out that the ordinary manifest worldview is presupposed in relations of mutual recognition. It is irrelevant whether mutual recognition can be modelled in some way with the resources of the scientific worldview—the participants must experience it from within their own perspectives. The idea that "recognition must be recognized" (explained in the second section below) settles the debate in favour of the manifest worldview, and the related understandings of the central concepts. The empirical

[1] See e.g. Searle (1995), Dennett (1995), Dreyfus (1992), Carroll (2015), Toivakainen (2015), Beran (2018), Gunkel (2012), Garber (2013), Enfield (2015), and Sandis & Harper (2017).

question of whether robots have the relevant capacities is to be understood in the way that the common-sense phenomenology, i.e. the manifest worldview, understands them, and not in the way the scientific worldview understands them. The paper ends with a brief look at whether mere simulated recognition, or merely apparent recognition, might serve to meet serve genuine human needs for recognition, and for example foster our self-esteem.

2 The Case: Mutual Recognition

This section explains the nature of mutual recognition, drawing on the literature during the last decades on the nature of mutual recognition. It describes the sense in which mutual recognition (respect, esteem, care) captures the normative grammar of human sociality (Sect. 2.1) and lists the ways in which recognition has been argued to matter (Sect. 2.2). It then explains the sense in which "recognition must be recognized", which turns out the be relevant in the next section (Sect. 2.3). The section further analyses the level of recognition preceding social recognition (esteem, respect or care), namely the attestation of capacities, expanding the theory of Paul Ricoeur (Sect. 2.4). It then asks what capacities an entity must have to be able to stand in relations of mutual recognition (Sect. 2.5).

2.1 Recognition as the Normative Grammar of Human Sociality

A putative theory concerning the "normative grammar" of human sociality is provided by the so-called theories of recognition.[2] Humans need recognition from others, need the experience that they count, or mean something, for the others. On this approach, how they matter to others can be analyzed in different dimensions of recognition, such as the need for being loved as an irreplaceable individual; the need for getting esteem, appreciation, admiration or gratitude for one's well-intended contributions and sacrifices for the sake of others, for doing one's share in the common good; and the justified claims for basic respect as someone possessing human dignity, and personal and

[2] Research on mutual recognition has been very lively during last decades. See e.g. van den Brink and Owen (2007), Deranty (2009), Fraser and Honneth (2003), Halbig and Quante (2003), Hegel (1977 [1807]), (1991 [1821]), Honneth (1995 [1992]), (2007 [2000]), (2008), (2002), Ikäheimo (2002, 2016), Ikäheimo and Laitinen (2007a, b, 2011), Iser (2013), Laitinen (2002), Markell (2003), McNay (2008), Pippin (2008), Schmidt am Busch and Zurn (2010), Seymour (2010), Siep (1979), Taylor (1992), and Williams (1998).

collective autonomy. Often these different forms of social recognition are summarized in three main kinds, respect, esteem and care, but nothing prevents us from distinguishing ten or more importantly different kinds of social recognition.[3]

A value-based approach holds that what makes respect, esteem or care a "fitting" or "appropriate" or "required" response are the evaluative features, which make someone esteem-worthy, calling for care or demanding of respect.[4] Other sorts of responses would amount to "misrecognition" or "non-recognition" rather than adequate recognition. Suffering from misrecognition is arguably a central factor in people's quality of life, and a source of social unrest. Adequate positive regard from others then supports positive relations to self: care of self, self-esteem and self-respect.[5]

The value-based approach thus sees the normative grammar of recognition between persons as a special case of the more general view that some aspects of the world are normative. Persons are special, but the speciality does not mean that we have normative expectations concerning persons only. A closely related conceptualization is that of "reactive attitudes"[6] such as blame and praise, moral indignation and resentment. When someone stamps on one's foot, there is a great difference in one's emotional reaction depending on whether the person did it on purpose or by mistake. There is a range of attitudes, then, that are properly directed only at persons, even if they are merely a special case of normative expectations (we may for example expect clocks to show time accurately, as that is how they ought to be like).

The theories of recognition and reactive attitudes typically stress that they demarcate between relations between human persons and the human relations to other things, which cannot be held immoral, irresponsible, or irrational. A hailstorm that destroys my garden is not to be held responsible for the deed, and I cannot blame my bicycle for its broken tire. It has been argued, perhaps in a too restrictive manner, that recognition is a "personifying" attitude, and that reactive attitudes are possible only with persons.[7] This is supported by the valid *mutuality-insight* concerning recognition: there can be mutual recognition only between parties that are recognizers. (There are also further implications, discussed in the next subsection).

[3] Laitinen and Pirhonen (2019).

[4] Laitinen (2002).

[5] Honneth (1995 [1992]).

[6] Strawson (1962).

[7] Strawson (1962) and Ikäheimo (2002, 2014).

A second insight, also supported here and pulling in a slightly different direction, is the *adequacy-insight*: what makes recognition appropriate is that it is an adequate response to the normative features (respect-worthiness, esteem-worthiness, vulnerability) of persons—analogously to responding adequately to the normative features of anything. On this approach, the adequate consideration need not be limited to features that are specific to persons. I may want others to take into account my capacity to feel pain, even though that capacity is not specific to persons. I feel misrecognized when that capacity is ignored.[8] Genuine "recognition" in the relevant sense then goes further than mere "identification" of something as something. It involves normative responsiveness in actions, behavioral dispositions, attitudes, expressions, and emotions.

2.2 Self-relations and Five Other Reasons Why Recognition Matters in General

A common starting point is the idea that getting recognition matters for the practical "relations–to–self" (such as self-respect and self-esteem) of the recognized ones. Respect is related to self-respect, esteem to self-esteem, denigrating feedback concerning one's abilities is related to an internalized sense of incapacity, experienced humiliations are related to a sense of inferiority.[9]

The connection between recognition from others and self–relations is a readily intelligible and interpretative one, despite also being causal. There is an intelligible, rational connection between "*A* thinks or feels such-and-such about B" and "*B* thinks or feels such-and-such about B". Now, various other things may also affect relations-to-self. Say, eating pills with certain chemicals may cause changes in one's moods or brain functionings, which may then causally contribute to changes in one's self-esteem. These kinds of effects are totally different from the way in which, say, respect from others is connected intelligibly to self-respect. They have none of the rational connection between A's and B's thoughts about B. It seems that recognition from others matters to relations-to-self in a distinctive "readily intelligible" way, which we understand when we understand what recognition is. Such relevance to relations–to–self may even be central in trying to delineate what recognition is: perhaps any feedback from others, which is relevant for one's relations–to–self in the readily intelligible way, will indeed count as recognition. This insight can be

[8] Laitinen (2010).

[9] See Taylor (1992), Honneth (1995 [1992]) and Walker (2006).

developed in two rival directions, which are in tension (one stressing mutuality, the other stressing adequate regard; as mentioned above).[10] In addition to the relevance to relations–to–self there are at least five further (related) ways in which getting recognition matters. It is good to mention these to prevent the impression that relevance to positive or negative self-relations is the only reason why recognition matters. A plausible definition of recognition should leave conceptual room for all these ways in which recognition matters, and a satisfactory full account of recognition would either cash out all these claims or explain why some of them are mistaken.

First of all, recognition may well be *directly desirable* in itself. Desire for recognition in its different forms is an intelligible independent motivational force. Note that a desire for recognition can be abused by others—say, by giving merely symbolic recognition to underpaid workers. But its existence can also be falsely neglected—as in the economistic assumption that the only thing that motivates people is money and economic rewards. It may well be, like Aristotle thought, that people are sometimes motivated by pleasure, sometimes by honour or recognition, and sometimes by virtuous goodness, but it is also important to notice that virtuous persons may desire—in addition to acting virtuously—also distinctive forms of pleasure, and they may also want adequate recognition (perhaps defined as responsiveness to goodness). Healthy pride comes with a reasonable expectation to be esteemed by reasonable others. Thus, a phenomenology of human action should account for one of the main springs of human motivation: a direct desire for recognition, which may or may not be useful in other ways, may or may not be pleasurable, and may or may not be consistent with virtue (such as the tyrannic desire to be recognized as superior to others).

Secondly, as mentioned, recognition matters to the positive relations-to-self, which again may be something that matter in their own right. But via affecting self-relations of the relevant parties, getting recognition may at the same time be a *precondition of human agency* (Honneth 1995 [1992], Anderson and Honneth 2005). As Anderson and Honneth sum it up:

> In a nutshell, the central idea is that the *agentic competencies* that comprise autonomy require that one be able to sustain certain attitudes toward oneself (in particular, self-trust, self-respect, and self-esteem) and that these affectively laden self-conceptions – or, to use the Hegelian language, "practical relations to

[10] The readily intelligible way in which recognition from others is relevant to the practical relations-to-self is especially helpful in guiding us in the question of what recognition is, or what all subspecies of recognition have in common. But it can lead in different directions when combined with an idea about adequate regard and when combined with an idea about mutuality.

> self" – are dependent, in turn, on the sustaining attitudes of others. (Anderson and Honneth 2005, p. 131; italics added)

So, sufficient recognition may be necessary for agency. Unlike animals guided by instincts, some agents may well be insecure without social support—too insecure to have the basic courage to act: it may be that the distinctive form of human agency is possible only when sufficiently recognized.[11] Although some forms of agency are possible without recognition, perhaps *responsible* or moral or rational agency presupposes recognition. In a related thesis, Hegel (1977 [1807]) claims that recognition is necessary for the kind of *self-consciousness* that human persons are capable of.[12]

If such features that are central to human personhood depend on recognition, we may come to appreciate *the ontological relevance of recognition to human personhood.* It might seem puzzling how recognition might play a role in the ontology of persons. But if we accept that human beings are *potential* persons when born, and acquiring the relevant capacities (agency, positive relations-to-self, self-consciousness, rational or moral agency) is dependent on recognition, there is no puzzle. Nor of course if we find acceptable that being recognized is directly one constitutive aspect of being a person.[13]

Thirdly, recognizing and getting recognition are directly constitutive of non-alienated horizontal *relationships of unity*, of different kinds (e.g., friendship, solidarity, mutual respect, mutual care), which are directly desirable, and are a matter of being at home (*bei sich*) in the social world. Recognizing and getting recognition are constitutive also of non-alienated *vertical* relationships of different kinds (e.g., living under just, legitimate, self-governed institutions, living under institutions whose goals and principles one can identify with), which are directly desirable, and a matter of being at home (*bei sich*) in the institutional world. Consider, for example, recognition from the state. Just like recognition from other individuals is relevant for the self–relation of the individuals, so is recognition from political institutions. And just like recognition from others is constitutive of reconciliation, unity or "being by oneself in another", so is recognition from the state. Or negatively, lack of recognition from the state may be constitutive of *alienation*, domination, and

[11] Cf. also MacIntyre (1999) and Honneth (2008).

[12] I will here rely on the interpersonal, "orthodox" readings of Hegel's *Phenomenology*, where the relations of recognition hold between two consciousnesses. The "heterodox" readings (e.g. Stekeler-Weithofer, McDowell) claim that at this stage, Hegel still talks about one agent. I thank Thomas Spiegel for commenting on this.

[13] For the debate on the ontological relevance of social relations, especially of relations of recognition, for human personhood, see the essays in Ikäheimo and Laitinen (2007b).

frustration. Some violations make the state something like an "enemy", some policies or goals make it more like a "stranger" and some structural features may make it "impotent" in crucial respects, and such experiences may lessen one's feeling of "being at home" in the institutional world. Recognition is thus directly constitutive of various social relations. Relatedly, we can point out the potential ontological relevance for the very existence of groups, institutions, or states. Independent states exist only when recognized by other states. It may be that the way in which groups and institutions are dependent on subjective takes can best be cashed out in terms of recognition.[14] Thus, we can point to the social ontological relevance of recognition, to relationships, social unities, groups and institutions.

Fourthly, recognizing and getting recognition is arguably in different ways a precondition of such desirable ends of individuals as identity-formation, self-realization, a good life, and positive freedom. For example, defining one's practical identity and fundamental aims in life is arguably a dialogical business, and self–realization via such identity-defining aims benefits from feedback and "personalized esteem" from others. Further, good life is arguably spent in a subjectively rewarding pursuit of objectively worthwhile activities and relationships, which include standing in relations of recognition, and arguably one's freedom is partly constituted via guarantees of non-domination which are forms of recognizing and respecting one's standing.[15] So, many desirable features of human life (well-being or good life, freedom or autonomy, authentic self-realization) may depend on recognition.

Fifthly, recognition may be equally relevant to various collective and societal ends. Various theories of justice, equality, solidarity, democracy or legitimacy, work or basic income, or even environmental philosophy and social robotics have in recent years put forward claims that the notion in question cannot be fully understood without recognition. Perhaps "recognition" is an irreducible aspect of justice, alongside distributive and participatory justice, or perhaps it is a good to be distributed among others, or a way of organizing participatory justice. And perhaps forms of solidarity can be cashed out in terms of mutual recognition.[16]

So, getting recognition matters in various ways: in addition to being desired for no further purpose, it can be relevant for both the existence of individuals and social groups, and various ethical and political ideals from autonomous good life to justice.

[14] Ikäheimo and Laitinen (2011).

[15] Laitinen (2022b).

[16] Laitinen (2022a).

2.3 What Is Mutual Recognition, and Why "Recognition Must Be Recognized"

This subsection details the structure of mutual recognition and explains why it is that recognition must be recognized. To get recognition from someone else, one must notice the other, recognize the other as a recognizer, and notice the act or expression that carries the recognitive message. Even in cases where B is otherwise being recognized by A, but B does not regard A as the type of entity that can give recognition, then B does not "get" recognition in the relevant sense (it does not affect B's practical relations-to-self, for example).[17] This may be the case if B is bigoted and thinks that women or people outside my tribe simply are not capable of giving recognition.[18] Or suppose B mistakes A for being so seriously mentally dysfunctional as to be incapable of forming attitudes about B. Or suppose B does not realize that A is using Skype, Zoom, or other forms of telepresence and B thinks they are just seeing a pre-recorded movie on the screen. I will use the slogan "recognition must be recognized" for this wide array of conditions for successfully getting recognition. It will be relevant for my argument below.

The general form of *mutual* recognition is: *mutual recognition is a matter of two recognizers mutually taking each other as recognizers of some kind and accepting the normative implications of such takings.*

Some clarifications are in order. First, this is an analysis of *mutual* recognition. Mutuality, in its *symmetrical* form, demands four things:

1. A takes B as R
2. B takes A as R
3. A takes A as R
4. B takes B as R.

That is, both take each other to be "R", namely *recognizers,* for example free and equal citizens, or human persons, or states. This is the outcome in Hegel's

[17] Laitinen (2010).

[18] In which case people of those sorts can perhaps engage in a struggle to be recognized as recognizers; but before they do, this kind of misperception may well prevail.

analysis in *Phenomenology of Spirit* and elsewhere.[19] Hegel stresses that self-relation and recognition are mediated by one another:[20]

> Each is the mediating term for the other, through which each mediates itself with itself and coincides with itself. Each is for itself and for the other an immediate self-existing being, which at the same time is such only through this mediation. They recognize themselves as reciprocally acknowledging each other." (Hegel 1977 [1807], §184).

Or as Ludwig Siep puts it:

> Recognition, as a double-signifying act of two self-consciousnesses, is a relation in which the relata relate to themselves through the relation to the other, and relate to the other through their own self-relation. Thus, the self's relation to itself is made possible by the corresponding relation to the other. (Siep 1979, p. 137)[21]

Second, this suggested definition or characterization concerns recognizers in general, whether they are individual human beings, groups or states, or indeed animals or robots. There are interesting questions concerning the capability of any of these to be a recognizer. Humans are not recognizers when they are born, and completely disintegrated groups or states cannot have shared attitudes or joint actions that it takes for them to count as agents of recognition. Things are complicated by the fact that recognition from others seems to play a significant role in the development processes where human persons learn to be agents of recognition themselves.[22] The same goes for the process where a less integrated group of people turns into well integrated one capable of having the needed attitudes, or where a state-candidate turns into an independent state. It seems that relations of recognition between the citizens, and the recognition that the citizens give to the state for its legitimacy, and

---

[19] This kind of mutual recognition of each other as recognizers can take place not only between two individual self-consciousnesses, but in principle between any two recognizers, for example between a state and an individual. The state considers itself to be a recognizer, it assumes it has the legitimate power to grant citizenship. The state recognizes the individual by granting it citizenship and rights. The individual takes the state to be a recognizer and thinks that something actually happened when the state granted citizenship, unlike if it was his neighbour's child who wrote the certificate. The individual takes herself to be a recognizer and thinks that the normative power of the state to recognize is ultimately based on its legitimacy among the citizenry.

[20] Hegel (1977 [1807]),§182, §184.

[21] Cited and translated in Williams (1998, p. 51).

[22] Laitinen (2002, 2010).

recognition received from other states are all different relations of recognition, which are necessary for a state to exist as an agent of recognition.

Thirdly, the normative implications in question depend on whether we have a case of interpersonal recognition between two persons, or international recognition between two states or peoples (e.g., concerning their independence and the right to govern themselves and status as a self-governing political whole), or perhaps between two, say, cultural groups. In addition to such horizontal cases, we have vertical cases, between an individual and a state, or between a state and an international organization like United Nations, or between a state and a minority group. Adequate recognition and misrecognition differ in how well the normative implications are taken into account and responded to.

These conditions are not sufficiently detailed, as they could be true by accident, as it were, without being interconnected in any way. As such, the four conditions allow for cases in which A and B have never met, and have only heard of each other on the radio, and they have no idea that they have heard of each other on the radio. Maybe this is quite acceptable, certainly this is not a case of *misrecognition*, and maybe such "disconnected" forms do count as recognition, although they cannot have effects of self-relations, because the recognizer is not aware of them. But perhaps we do not want such cases to count as "getting recognition" at all, once we analyze paradigm cases of full-fledged mutual recognition (which need not be a matter of face-to-face encounters, but which it is easiest to imagine that way), which is more demanding in terms of the kind of mutuality in question.

These more demanding, "connected" forms of mutuality can be illuminated when we focus on cases which are not strictly symmetrical over and above the minimum that both take each other to be recognizers. Take a case where an art-critic takes a guitar-player to be an excellent artist. This is not symmetrical: the guitar-player need not in return take the art-critic to be an excellent *artist*. But it is still mutual in another sense. What matters here is whether the guitar-player takes the art-critic to be a good art-critic, and thus a relevant and competent judge in this issue. If the critic understood the artist in the light that was intended and made good points, then his criticism cannot be just brushed aside. Comments from more competent and powerful judges and more significant others matter more. In the paradigmatic cases, recognition is relevant to self-relations.

In a less paradigmatic case, if A takes B to be an excellent guitar-player, but B does not know of this, or does not care about A's views (because he does not take A's judgements as competent or relevant in general), or does not think that this particular judgement is well-founded (say, the critic did not realize the ironic allusions, which were the point of the piece), or is not centrally concerned about what other people say about his guitar-playing (say, because he identifies himself more with his professional role), it may be that this criticism does not affect the self-relation of B much, if at all.[23] In general, the attitudes of B determine whether A *relevantly* and *successfully* takes B as X. Some of these attitudes are: B knows of A, B thinks of A as a competent judge in this case and B knows of and understands and accepts A's judgement that "B is X", and that it matters to B what others think of her X-hood. This also implies the minimal mutuality that both take both to be recognizers in general.

Can we say that these features that figure in "successful" or "relevant" recognition, are necessary conditions of recognition? With some qualifications, we can say that every form of recognition necessarily presupposes that both take each other to be recognizers. The qualifications concern potential persons (or potential recognizers in general) and imaginable cases, where there is some kind of esteem and some kind of love, but yet no judgement concerning the status of the other as a recognizer. What about the cases where B does not take A to be generally a competent or relevant judge in the issue at hand? Are they cases of recognition at all? Recognition matters more if it comes from a competent and relevant judge, but competence and relevance seem to come in degrees, so these conditions are better taken as qualifications of the recognition in question than as necessary conditions. It may be for example that B has no view at all concerning A's competence. Possibly even quite irrelevant recognition is just that, recognition but irrelevant.

B's self-relations depend on recognition from many persons, including recognition from A among others, and including this particular claim by A, among other claims. Here, a holistic approach is in place. Single judgements are always interpreted against the background of one's general sense of self and one's sense of how others view me. Sometimes single comments from others may have very crucial effects, often they are just drops in the ocean.

The same thing can be said of the question whether it matters to B what others say of her X-hood, for example, that it matters less to me what others think of me as an occasional dancer than what others think of me as a philosopher. Again, such mattering comes in degrees, and even irrelevant recognition may be recognition.

What about the conditions that B in successful cases *is aware of* A's claim that "B is X", and *understands* it and further that B *accepts* it, perhaps because he or she *agrees* with the judgement (or at least finds it reasonable)? These

[23] Laitinen (2003).

conditions are attempts to analyse what "successfully" means. It seems that awareness is a necessary condition for recognition to take place, but understanding comes in degrees. Further, it seems that something can be a case of recognition even where the recognizer herself does not agree with the judgement. Indeed, often the views of competent and relevant judges are all the more relevant when they differ from the recipient's own views.

2.4 Recognition as Attestation of Agential Capacities: In Oneself and another

Another set of distinctions will further be relevant here. Paul Ricoeur (2005) distinguishes between three senses of recognition: mere "identification" of anything, self-recognition as "attestation" to one's capacities, and finally, mutual recognition, *Anerkennung*, as discussed above. The topic of attestation is central to the difference between our relations to humans and to robots.

Ricoeur emphasizes an important presupposition of the debates about mutual recognition: the fact that we are able to act, that we are *capable agents*, and therefore capable of taking responsibility. This links up with Ricoeur's (1992) earlier analysis of the kind of certitude with which we recognize that we have various capacities as agents. It differs from mere descriptive "identification" of something as something. The type of recognition in question is "attestation", expressed by self-assertions such as "*I believe that I can*", and implicit in anything that we do. Ricoeur points out that there is a special sense of certitude or suspicion at stake in expressions of the type "I believe that I can". Ricoeur calls this "recognition-attestation" (Ricoeur 2005, p. 92). What is at stake is a kind of existential or practical confidence in one's capacities, which is not reducible to the degree of belief that the availed evidence would support. Admittedly, at any moment a sceptical self-doubt that I might have just become paralyzed and am no longer a functioning bodily agent is in theory possible, but more importantly there are more existential worries along the lines that "should I commit myself to this or that task – will I be able to cope?". The immediate object of such recognition-attestation is "capacities", but importantly my own capacities. The self in self-designation recognizes itself as the agent possessing these capacities. So ultimately, what is at stake is literally self-recognition: recognition of oneself by oneself.

Using the schema "A recognizes B as Z" it may seem that in Ricoeur's (2005) view in social *Anerkennung* ("being recognized by others") A and B are different persons and Z includes such aspects as respect for rights, esteem for achievements, love for the near and dear, and gratitude for gifts; whereas in the mode of *attestation* (self-recognition) A and B are the same person, and Z includes such central human capacities as to speak, to act, to narrate, to take ethical responsibility (as also analysed in *Oneself as Another*). One observation is that *Anerkennung* seems to concern social relations, social world, social roles (typical concerns of social philosophy), whereas *Attestation* concerns capacities of persons (that are presupposed in such social relations perhaps) (typical concerns of philosophical anthropology). Yet there is no reason to regiment *Anerkennung* and *Attestation* so strictly.

My claim is that *both* in social *Anerkennung* (respect, esteem, love, etc.) *and* in the attestation of human capacities all four combinations are important: recognition of self by oneself (A by A) and recognition of self by another (A by B), recognition of another by oneself (B by A) and recognition of another by him- or herself (B by B). They all are noteworthy possibilities, that Ricoeur just bypasses. It is the Z (and the appropriate mode or recognizing someone as Z) that goes to explain the difference between "attestation of human capabilities," and "respect," "esteem," and "love". Both recognition and self-recognition can be at stake concerning both.

Here is a further thesis: B can experience being respected, esteemed, or loved by A only if one can grant the A the status of being a recogniser. And one can grant that status only if one can grant this kind of "recognition-attestation" that A has the relevant capacities. If I cannot regard say a flower or a tree as having the relevant capacities, I cannot interpret the flower or a tree as giving me recognition. Whether we can honestly think that groups, non-human animals, or robots can have such capacities is relevant for our capacity to stand with them in relations of mutual recognition. Currently, it seems, we typically do not hold robots as having those capacities. That the worry of "deception" of therapeutic robo-seals like *Paro* is typically discussed in the case of demented elderly suggests that non-demented adults are not typically regarded as regarding robots as having the relevant capacities.

2.5 What Capacities Does It Take to Be a Recognizer?

Interpersonal recognition can take place only between persons. So one answer to the question of whether mutual interpersonal recognition is possible between A and B is given by the capacities required for being a person: are A and B persons? In fact, the capacity for mutual recognition is one of the six standard conditions of being a person listed by Daniel Dennett.[24] While we

[24] It would deserve further study how these capacities are interrelated. For example, self-consciousness does not as such suffice for the capacity to give recognition, as recognition is not mere "identification" but requires a practical, normative responsiveness, see the discussion above.

need to operate in the "intentional stance" in interaction with robots, fully "personifying" stance would arguably be a category mistake (Dennett 1976, 1987). Social robots form perhaps a new category, new vaguely demarcated "social grammar", albeit one with genuine normative expectations and experiences. Rewarding experiences in responses from robots need not be deceptive, although taking a fully "personifying stance" would be deceptive. Dennett (1987) suggests we can take different stances towards a chess computer. A natural scientist may see it as particles, the engineer in terms of flow charts and functions, but a chess-player will approach it with an intentional stance—as an opponent, which makes moves, plans them ahead and so on. Clearly all sorts of robots can be approached with the intentional stance. The "personal stance", or what I call here "personifying stance", is according to Dennett (1976) appropriate when dealing with things which meet the "conditions of personhood", such as rationality, capacity to reciprocate, language, self-consciousness:

1. Persons are *rational beings*.
2. Persons are beings to which states of consciousness are attributed, or to which psychological or mental or *intentional predicates* are ascribed.
3. Whether something counts as a person depends in some way on an attitude taken toward it, a *stance adopted* with respect to it.
4. The object toward which this personal stance is taken must be capable of *reciprocating* in some way.
5. Persons must be capable of *verbal communication*.
6. Persons are distinguishable from other entities by being *conscious* in some special way: there is a way in which we are conscious in which no other species is conscious. Sometimes this is identified as *self*-consciousness of one sort or another.

More should be said about what sort of capacities regarding reciprocation. It includes the capacity to form the basic stance or attitude towards the other, understanding the other as an individual that is behind all the expressions and activities. This presupposes the capacity to form beliefs, to be motivated, to have emotions (necessary for reciprocating emotions), be capable of empathy (necessary for understanding emotions and basic attitudes), and as adequate recognition includes *normative* responses, reasons-responsiveness or rationality as well as moral responsibility. Further, the capacity (4) seems partly already to presuppose (5) as recognition must show in verbal and bodily communication, and on some views, it presupposes (6). On other views, reciprocating may precede full self-consciousness.

3 Do Robots Have Such Capacity to Be Recognizers?

In this section I will address the surprisingly deep disagreement on the apparently simple empirical question whether robots have the capacity to be recognizers. And even though it seems like an empirical matter, it goes deep in to rival approaches to what we mean by the central concepts, illustrated by the debate between the likes of Dreyfus (1992) and Searle (1980, 1995) on the one hand, and Dennett (1995) on the other hand.

In 1950, Alan Turing discussed (while not subscribing to it) the line of argument that machines will never learn to:

> ...be kind, resourceful, beautiful, friendly, have initiative, have a sense of humour, tell right from wrong, make mistakes, fall in love, enjoy strawberries and cream, make some one fall in love with it, learn from experience, use words properly, be the subject of its own thought, have as much diversity of behaviour as a man, do something really new."(Turing 1950, p. 447)

Interestingly, Turing thematizes the different uses of words when he writes:

> The original question, 'Can machines think?' I believe to be too meaningless to deserve discussion. Nevertheless, I believe that at the end of the century *the use of words and general educated opinion will have altered so much that one will be able to speak of machines thinking without expecting to be contradicted.*(Turing 1950, p. 442)

Turing thus predicted that the use of words and educated opinion would change so that everyone uniformly would take it meaningful to speak about machines thinking. What has happened instead—and this is my first interpretation of the current predicament, – is that it seems that the whole list of features that distinguish human persons from robots suffers from a double talk: same central words are used in two systematically different ways. For all of the features, one can give an (impoverished) sense in which technical systems have them already today, and a more full-fledged, perhaps genuine, sense in which they do not. So, when there seemingly is a debate on whether robots can think, the parties may be talking past each other—it may be that both parties in fact agree that robots do not in some sufficiently inflated sense Think, Believe etc. and that in some sufficiently deflated sense they can "think". The disagreement is whether the question "can machines think?" is to be interpreted as "can machines Think?" (to which the answer on all sides may be "no"), or as "can machines 'think'?" (to which the answer on all sides may be "yes").

My second interpretative claim is that on the approach that Sellars would call the "manifest image" and I will call "common-sense phenomenology", the question is to be interpreted in the former way (Sect. 3.1), even if on some "scientific" worldviews, for example that of cybernetic cognitive science, the question can be interpreted in the latter way (Sect. 3.2).[25] There is systematicity to the double talk. In the manifest image (common-sense phenomenology), humans *think and act*, robots merely *"think" and "act"* in some deflated sense. There is the normal usage versus mere simulation. For the scientific approach, folk psychology seems to wrongly assume that humans would need to *Think and Act* in the inflated sense; whereas for the scientific approach, both humans and robots *think and act* (when these are understood in sufficiently deflated sense). Both perspectives think they are entitled to the normal use and have a pejorative view about their opponents: from within the common-sense view, the rival usage settles too cheaply for mere deflated "thinking" and "acting" (and is right that both robots and humans are capable of that), and from within the scientific perspective, the common-sense usage hankers after overly inflated Thinking and Acting (and while the common sense is right that robots are not capable of that—why should we accept that humans are capable of it either?). So there are systematically three readings: inflated Thinking, ordinary thinking (which the two viewpoints genuinely interpret differently), and deflated "thinking".

In the context of such systematic double talk, the question of whether humans and robots can stand in relations of mutual recognition will then depend on which of these worldviews we have adopted for posing the question (Sect. 3.3).

[25] Sellars's (1962) distinction between manifest and scientific image resembles many similar distinctions: Charles Taylor (1985) distinguishes between an engaged and a disengaged view, Thomas Nagel (1986) distinguishes a view from within and a view from nowhere, John McDowell (1996) distinguishes a view from within with sideways-on picture, Peter Strawson (1962) distinguishes a participant and an observer standpoint; and Bernard Williams (1978) between a subject-dependent versus "absolute" description of the world. Should we just drop the disengaged perspectives and opt for a direct engagement within the manifest worldview?. Paul Ricoeur (1992) favors indirect hermeneutics instead: a detour via various objectivist analyses and return to a "second naivité". From the engaged picture via several objectivist, scientific images to a fragmented big picture, which allows no synthesis, but retains the various "images" of the same reality.

3.1 Common Sense Phenomenology and the Manifest Worldview

Here is a line of reasoning from what I will call the standpoint of "common-sense phenomenology": Persons have, but stones and mindless machines don't have the following: first, an interlinked network of psychological features: sentience, feeling, emotion, desire, will, intention, commitment, belief, thinking, caring, acting, reasoning, communicating. Secondly, a world, a life, a lived body (*Leib*) in addition to physical body (*Körper*). Third, capacities for consciousness, sentience, self-consciousness, self-determination, moral and responsible agency. Fourth, the capacities to self-attest to one's capacities, the capacity to have expectations about how others think they ought to be treated and experiences of misrecognition when they are not, the capacity to acknowledge normative entities- and thereby, capacities to stand in relations of mutual recognition with other persons.

Robots can perhaps simulate these, but not have the genuine ones. They cannot intend or care or believe, but perhaps they can "intend", "care", "believe" etc. The same for the full list of capacities that humans have, and robots do not. (Cf. Seibt et al. 2016 on five types of simulation).

It is further important that the features come in holistic networks: robots cannot have desires or intentions, unless they also have beliefs. For them to acquire one capability from a network, they must acquire the rest (or at least some) of them. They can perhaps be had in rudimentary forms, but they cannot be had one by one. (cf. Davidson for one defense of such holism).

Further, there is a qualitative leap to be made by future robots, before they are sentient, intending, self-conscious etc. beings like humans, instead of merely simulating them.

From this standpoint robots fail to meet the criteria of personhood and the relevant capacities of recognizers. This standpoint assumes that social robots lack now—and will lack in the future unless a qualitative leap takes place—the phenomenal aspect of consciousness: there's nothing that it feels like to be a robot. They do not feel emotions, although they may simulate expressions of emotions in speech and behaviour. Thereby they will lack direct primitive self-consciousness, although they may have higher-order states that represent their own states. (By contrast, "cyborgic" extensions of humans will have the human consciousness and living body as their starting point, so they will provide a different kind of case).

Although they will have highly advanced and intelligent "system 2" they arguably do not have any kind of "system 1", as they are termed in dual process theories: system 1 is fast, automatic, emotional, intuitive, whereas system 2 is slow, effortful, logical, calculating and conscious.[26] *They* do not get upset, angry, emotional, feel pangs of conscience, guilt, shame, pity, compassion, sympathy, empathy even though they can simulate expressions of these. Social robots can perhaps read people's emotional states, or sense their insecurity, but they cannot emote. Robots themselves do not have a "manifest worldview", they only have a "scientific worldview".[27] They do not have *a world* (a lifeworld of perceived significances), only an environment.[28] They do not read or target emotions as such, but they rather track signs and expressions of emotions. Let us agree then that this amounts to them not having the status of persons. It will remain acceptable to, say, turn them off or disentangle them into pieces—which would be grave violation of human dignity when done to humans.

As has often been pointed out, there may be an element of deceptiveness if robots behave as if they have the "system 1", say, as if they would have genuine feelings or genuinely care. Approaching such robots with the full blown "personal stance" will contain an amount of deception, unless they actually do have such a system 1. If robots are one day to come to have genuine emotions, thoughts, intentions, attitudes, to have a world, to be conscious or self-conscious, it takes a qualitative leap in the future—currently they lack them altogether.

This leads to an interesting conclusion: If this common-sense approach gets things right, humans and robots cannot stand in relations of recognition, because robots lack all the relevant capacities. They are not sentient or cannot suffer, so we cannot care about them; and as they cannot think, emote or form attitudes, we cannot be insulted or disrespected by their attitudes. They do not have the capacities that would make them persons, and so interpersonal recognition is out of the picture. They do not have the relevant capacities for giving or receiving recognition.

In the future robots may acquire the relevant capacities. But interestingly, because "recognition must be recognized", it means that even if this common-sense approach would get things in fact wrong, but is the common-sense

---

[26] Kahneman (2003, 2011).

[27] Sellars (1962). Of course, the claim that robots cannot emote is open to the same dispute between worldviews as the clam they cannot think: can they Emote, emote, "emote"? The argument here is that from the manifest worldview, robots do not emote, but at most "emote" in some deflated sense—that there is a significant difference between humans and robots, from the viewpoint of common-sense phenomenology. The argument then turns on the claim that for certain phenomena, like mutual recognition, we must adopt the lived viewpoint of common-sense phenomenology. I thank Thomas Spiegel for commenting on this.

[28] cf. the Movie *Matrix,* 1 s and 0 s instead of phenomenal experiences.

worldview of ordinary people, then ordinary people cannot get recognition from robots, because they do not regard robots as recognizers.[29] Because of this, it may be that for social robots to really succeed in replacing humans in relations of mutual recognition, it would take persuading on humans in their everyday lives to drop their everyday "manifest worldview" and take the scientific stance. That may well be impossible given the nature and point of the common-sense worldview—as it may well be necessary for participation in ordinary life.

3.2 Scientific Worldview and the "Cybernetic" Approach

The scientific worldview has made progress by questioning various "folk" assumptions, and putting aside various meanings that guide ordinary people in their lives. For example, a "cybernetic" approach to humans and machines as cognitive systems would proceed on the basis of a scientific worldview. On this approach, no qualitative leap is needed for humans and robots to be in the relevant senses similar. No phenomenal consciousness is needed (*pace* the erroneous common-sense expectations). Humans and robots are ultimately information processing systems, and certainly robots, AI, or computers will be better than humans in information processing in various respects. Folk psychology may assume that we need "phenomenal consciousness" to explain outward behavior; but do we? Don't we have better scientific psychology so that the folk psychology can be just put aside?

The double talk comes to fore when we note that cybernetic systems *in a sense* are "conscious", they have representations of their environment they can access (access consciousness); and they have representations of goals (that function like intentions); and they have representations of their own states (that functions like self-consciousness). In a sense, such systems distinguish between themselves and the environment (i.e. in a sense have a "world"). Further, when the robot is turned on, it not only has a physical body—and while it admittedly does not have a lived body, it has a functioning processual artefactual body irreducible to mere physics.[30] It "knows" the position of its limbs and can use its body in acting. Perhaps it doesn't consciously Know, or

[29] In such a case, humans could be to robots like Hegelian masters who do not yet regard their slaves or servants as capable recognizers, or like paranoid tyrants who cannot believe their oppressed and fearful subjects to be honest or indeed capable of forming thoughts at all, or like bigots who cannot extend the status of a recognize to apparent people outside one's tribe (Hegel 1977 [1807]).

[30] Cf. Dennett (1987).

freely Act, but *so what?* the cybernetic approach asks. So, why hanker after Consciousness, Self-Consciousness, Reason, Action, or Morality on the list, why not stick with the kind of consciousness, self-consciousness, reason, action, or morality that machines *can* have?

3.3 Is a Stereoscropic View Possible in this Case?

In general, one can see the task of philosophy as uniting the perspectives, in a stereoscopic image, as Sellars would say. In doing so, it is good to see what the rival approaches are good for. The manifest worldview is not primarily needed in the business of scientific "explaining", but in "living". The lived experiences are not explanatory assumptions or folk theories; and the common-sense approach is basically right about the nature of the manifest world. The scientific worldview may be basically right about what assumptions are needed in the game of providing explanations—they may be right to criticize "folk psychology" *understood as explanatory assumptions*. So, we need (at least) a stereoscopic vision which does not reduce either image to the other; we should adopt Sellars's task (without ending up with his scientism).

But for the specific question at hand, the manifest worldview and the purposes of "living" are central. Humans need recognition in their ordinary lives, to even become capable of adopting the scientific stance.[31] The relations of recognition must be antecedent to such highly sophisticated cognitive stances (see e.g. Honneth 2008). This may be what lies behind the widely spread suspicion that we should not leave our children to be raised by robots assuming their needs for mutual recognition would not be adequately met. Further, because, in a slogan, *recognition needs to be recognized* — for example because recognition must be experienced as taking place if it is to affect our relations-to-self —, it is neither here nor there if in a scientific model we can model robots as having the network of capabilities of recognition in some deflated sense. If humans need recognition which is recognizable as such from within the manifest worldview, we can conclude that *in this case* we should give priority to the manifest worldview, and accept that robots cannot think, act or be self-conscious in the relevant senses. They can only "think", "act" or be "self-conscious". The best that robots can provide is simulated recognition.

Here a possible objection must be addressed: if there is such systematic double talk as claimed above, and people's understandings about the central concepts are diffuse, is it not simply wrong to state that people's

[31] For an argument examining what this means for normativity of the scientific and manifest worldviews, see Spiegel (2022).

understandings are univocally on the side of the "common sense phenomenology"? And is not the scenario where, for example, a demented elderly person regards robo-seals as genuine recognizers a genuine and widely discussed threat? With technological advancement, will not that be the case with also non-demented adults? Here I think it must be admitted that such technological advancement is certainly imaginable, but it also tells us something that the worry nowadays concerns precisely the demented individuals. Further, given the diffuse understandings we must draw a distinction between people's first impressions, and their reflected, clarified judgements about what categories they inevitably use in their life. The claim is that any ordinary person, with a suitable period of sustained reflection, will come to broadly similar views concerning what it takes for something to be responsible agent, or an entity capable of recognition, and that they would regard their laptops as not responsible agents or agents capable of recognition in the relevant sense. The claim is not that everyone already agrees with this philosophical insight, but that if they would think through these two different options (on the question of whether machines can think in the relevant sense, so that mutual recognition is possible with them), once the conceptual distinctions have been clarified, they would in their considered opinions come to appreciate the philosophical insight that robots do not have the required capacities in the relevant sense (even though they have them in some deflated sense, which does not suffice for genuine recognition). But even more importantly, they cannot give up the categorizations of the manifest worldview in their everyday life and those categorizations lead (when scrutinized carefully) to a sharp difference between humans and robots.

4 Could Simulated Recognition Be Enough? Could Recognition from a Robot Enhance One's Self-esteem?

This final subsection asks whether human needs for recognition could to some extent be met even though recognition comes from what is non-deceptively regarded a mindless machine. Arguably, say, access to television and movies, even though not genuine interaction, can provide some sense of human company. If deserted on an island (with electricity available) many of us would prefer having a library of movies and Blu-Rays, and a television set, to not having them. I take it for granted that we would not want to replace primary caregivers with robots, to the extent that we want human infants to experience genuine encounters with genuinely emoting and communicating humans. If humans need recognition to develop their capacities, we can argue that in the first stages of life, humans need genuine recognition from other humans. This is allegedly stronger than the need for recognition in any other phase of life (perhaps at old age loneliness is a distinct harm, but at that age recognition does not anymore have the function of helping us become full persons).But perhaps there is a limited sense in which automated systems could enhance our practical relations to self in our adult lives, for example our self-esteem.

In *work*, or *practical activity* more generally, we get feedback from the material reality: if we try to fell a tree, when the tree indeed falls, that is part of the basis on which we can feel satisfied for having succeeded. This is one way in which feedback from machines, artificial intelligence, can matter: if I try to beat a chess machine, and succeed, then I get objective feedback, which is in principle non-social. When a human person greets me with joy, I'm happy that he or she not only identifies me but is further happy to see me. When my pet dog greets me with joy, I'm happy that she not only identifies me but seems happy to see me. Thus, in some sense we seem to regard even pet dogs as relevant parties in that they give social recognition for us that we experience as rewarding.[32] What about the following: When I drive my car to a parking hall with automated licence plate recognition system, and the gate opens without me needing to press a button, I'm recognized by the system in the sense that it *identifies* me. I probably will not feel "getting recognition" by thinking that the parking hall is somehow happy to see me; that would be somewhat exaggerated. But I may feel that I get recognition in some minimal sense, of being someone in the human social world, of having the status of being allowed to drive in; and even able to drive in without pressing the button.

Reliable automated responses to a human-generated status (identified via the licence plate in this case) can feel like "getting recognition", getting confirmation to my status and standing. I need not deceive myself into thinking that the parking hall gate is the recognizer, rather the automated system "sends a message" in an analogous way to how practices of legal punishment "send a message".[33] What messages the system sends depend on what the planners and programmers have built in, but just like meanings in a text may exceed the authorial intention, the messages sent by a technological system may exceed the designers' intentions.[34] Say, although it was not intended to so do,

[32] Laitinen (2010).
[33] Hampton (1992).
[34] Ricoeur (1991).

a piece of technological equipment can send a message "you are too slow and your eye-sight is too bad" (as my smartphone repeatedly seems to do).

Consider the difference in the "message sent" in an interpersonal encounter, where one communicates with body language, gestures, choice of words and topics, tone of voice, that the other really matters to me. The message one gets is in the embodied being, in the expressions of the other, including conscious intentions. By contrast, in a habitual routine performance of a social practice, where it is the social practice as a whole that sends the message (say, being allowed to enter a barber shop, or not being allowed). In being treated in some manner, which is responsive to one's status, one *gets a message* from the social practice, even though that message is not in the personal attitudes of the role-occupier. The level of habits and practices is arguably no less important than the interpersonal one-to-one encounters. And as there are so significant differences in the "mindedness" of the origins of messages, there may well be automated "messages" without minds.

Now, the "messages sent" by automatons can also be relevant for one's self-esteem. Consider a society where gates automatically open to people of some kind, but not to people of some other kind. Such "messages sent" can be highly relevant in the social bases of self-relations. Again, some of the messages may be just like the designers and programmers intended them to be, but some messages may exceed the intentions. These messages sent by the automatons can be regarded as texts: reading a book, which denigrates me and my kind—even when anonymous—will be meaningfully experienced as genuine denigration even though it is delivered in a text. With written texts, it is transparent that it is not the book that originates the message, it merely delivers it.

The more autonomously generative the computer or robot is, the more it takes the role of the originator of the message. Suppose an algorithm makes suggestions based on our preferences, but also on the basis of some background info, say, what is suggested to women and to men. Various algorithms already follow our shopping behaviour or "liking behaviour" in Facebook. They could in principle be programmed to generate nice or insulting things to say to us—and they could learn and generate new insults that the engineers never programmed in. The more this happens, the closer the question is whether we'd regard such programmed robots as capable of not merely delivering recognitional messages from other parties (designers, engineers, society), but of authoring the messages themselves.

5 Conclusion

Do robots have the capacities and functionings to be able to stand in relations of mutual recognition with humans? Some say "obviously not" and some others "in the relevant sense, yes". This article has argued that given the nature of the manifest worldview, and the demand that recognition must be experienced for it to affect our relations-to-self, the correct answer is "in the relevant sense, no". And as they lack the requisite capacities, robots can at best simulate genuine recognition. The last section suggested briefly that such simulated recognition, known to be simulated, might nonetheless have some relevance for example for our self-esteem.[35]

References

Anderson, Joel, and Axel Honneth. 2005. Autonomy, vulnerability, recognition, and justice. In *Autonomy and the challenges to liberalism: New essays*, edited by John Christman and Joel Anderson. New York: Cambridge University Press.

Beran, Ondřej. 2018. An attitude towards an artificial soul? Responses to the 'Nazi Chatbot'. *Philosophical Investigations* 41 (1): 42–69.

Carroll, Sean. 2015. We are all machines that think. In *What to think about machines that think: Today's leading thinkers on the age of machine intelligence*, edited by John Brockman, 56–58. New York: Harper Perennial.

Dennett, Daniel. 1976. Conditions of personhood. In *The identities of persons*, edited by A. O. Rorty. University of California Press.

Dennett, Daniel. 1987. *The intentional stance*. Cambridge, MA: MIT Press.

Dennett, Daniel. December 21, 1995. The Mystery of Consciousness: An Exchange [with John Searle]. *New York Review of Books*. https://www.nybooks.com/articles/1995/12/21/the-mysteryof-consciousness-an-exchange/

Deranty, Jean-Philippe. 2009. *Beyond communication—A critical study of Axel Honneth's social philosophy*. Leiden: Brill.

Dreyfus, Hubert. 1992. *What computers still can't do*. MIT Press.

Enfield, N. J. 2015. Machines aren't into relationships. In John Brockman, ed. *What to think about machines that think: Today's leading thinkers on the age of machine intelligence*. Harper Perennial, New York, 397–398.

Fraser, Nancy, and Axel Honneth. 2003. *Redistribution or recognition? A political-philosophical exchange*. London/New York: Verso.

Garber, M. Sep 20, 2013. Funerals for Fallen Robots, *The Atlantic*.

[35] I would like to thank friends, colleagues, and workshop-participants with whom I have had a chance to discuss these topics over the years, as well as Rosalie Waelen for the kind invitation.

Gunkel, David J. 2012. *The machine question: Critical perspectives on AI, robots, and ethics*. Cambridge: The MIT Press.
Halbig, Christoph, and Michael Quante, eds. 2003. *Axel Honneth: Sozialphilosophie zwischen Ethik und Anerkennung*. Münster: LIT.
Hampton, Jean. 1992. Correcting harms versus righting wrongs: The goal of retribution. *University of California Los Angeles Law Review* 39:1659–1702.
Hegel, G. W. F. 1977. *[1807]. Phenomenology of Spirit*. Oxford: Clarendon Press.
Hegel, G. W. F. 1991. *[1821]. Elements of the philosophy of right*. Cambridge: Cambridge UP.
Honneth, Axel. 1995. *[1992]. The struggle for recognition*. Cambridge: Polity Press.
Honneth, Axel. 2002. Grounding recognition: A rejoinder to critical questions. *Inquiry* 45:4.
Honneth, Axel. 2007. *[2000]. Disrespect: The normative foundations of critical theory*. Cambridge: Polity Press.
Honneth, Axel. 2008. *Reification. A new look at an old idea*. Oxford: Oxford University Press.
Ikäheimo, Heikki. 2002. On the genus and species of recognition. *Inquiry* 45:4.
Ikäheimo, Heikki. 2014. *Anerkennung*. Berlin: De Gruyter.
Ikäheimo, Heikki, and Arto Laitinen. 2007a. Analysing recognition—Identification, acknowledgement and Recognitive attitudes between persons. In *Recognition and power*, edited by v.d. Brink & Owen. Cambridge: Cambridge University Press.
Ikäheimo, Heikki, and Arto Laitinen, eds. 2007b. *Dimensions of personhood*. Academic Imprint.
Ikäheimo, Heikki, and Arto Laitinen, eds. 2011. *Recognition and social ontology*. Leiden/Boston.
Iser, Mattias. 2013. "Recognition", *Stanford Encyclopedia of Philosophy*, http://plato.stanford.edu/entries/recognition/.
Kahneman, Daniel. 2003. A perspective on judgement and choice. *American Psychologist* 58:697–720.
Kahneman, Daniel. 2011. *Thinking—Fast and slow*. Macmillan.
Laitinen, Arto. 2002. Interpersonal recognition - a response to value or a precondition of personhood? *Inquiry* 45:4.
Laitinen, Arto. 2003. Social equality, recognition, and preconditions of good life. In *Social inequality today. Proceedings of the 1st annual conference of the CRSI 2003*, edited by Michael Fine, Paul Henman, and Nicholas Smith. North Ryde: CRSI, Macquarie University.
Laitinen, Arto. 2010. On the scope of 'recognition': The role of adequate regard and mutuality. In *Philosophy of recognition: Historical and contemporary perspectives*, edited by H.-C. Schmidt am Busch and C. Zurn, 319–342. Lanham: Rowman & Littlefield.
Laitinen, Arto. 2022a. Solidarity. In *Oxford Research Encyclopedia of Politics*. https://doi.org/10.1093/acrefore/9780190228637.013.2013.

Laitinen, Arto. 2022b. Mutual recognition and Well-being: What is it for relational selves to thrive? In *Theory and practice of recognition*, edited by Onni Hirvonen and Heikki Koskinen. Ch.3.

Laitinen, Arto, and Jari Pirhonen. 2019. Ten forms of recognition and misrecognition in long-term Care for Older People. *Sats* 20 (1): 53–78.

MacIntyre, Alasdair. 1999. *Dependent rational animals. Why human beings need the virtues*. London: Duckworth.

Markell, Patchen. 2003. *Bound by recognition*. Princeton: Princeton University Press.

McDowell, John. 1996. Mind and world. With a new introduction by the author. Harvard UP: Cambridge, Mass.

McNay, Lois. 2008. *Against recognition*. Cambridge: Polity Press.

Nagel, Thomas. 1986. The view from nowhere. New York: Oxford.

Pippin, Robert. 2008. *Hegel's practical philosophy. Rational agency as ethical life*. New York: Cambridge University Press.

Ricoeur, Paul. 1991. *From text to action: Essays in hermeneutics II*. Evanston: Northwestern University Press.

Ricoeur, Paul. 1992. *Oneself as another*. Chicago: The University of Chicago Press.

Ricoeur, Paul. 2005. The Course of recognition. Cambridge, Mass.: Harvard UP. Transl. David, Pellauer; orig. 2004. Parcours de la Reconnaissance. Paris: Éditions Stock.

Sandis, Constantine, and R. Harper. 2017. The AI of the beholder, https://medium.com/@constantinesandis/the-ai-of-the-beholder-303ecac5fc9d.

Schmidt am Busch, Hans-Christoph, and Christopher Zurn, eds. 2010. *The philosophy of recognition*. Lanham: Lexington Books.

Searle, John. 1980. Minds, brains and programs. *Behavioral and Brain Sciences* 3:417–457.

Searle, John. December 21, 1995. The Mystery of Consciousness: An Exchange. A Reply [to Dennett]. New York Review of Books.www.nybooks.com/articles/1995/12/21/the-mystery-of-consciousness-an-exchange/

Seibt, Johanna, Marco Nørskov, and S. S. Andersen, eds. 2016. *What social robots can and should do*. Amsterdam: IOS Press.

Sellars, Wilfrid. 1962. Philosophy and the scientific image of man. In *Frontiers of science and philosophy*, edited by Robert Colodny, 35–78. Pittsburgh, PA: University of Pittsburgh Press.

Seymour, Michel, ed. 2010. *The plural states of recognition*. Hampshire: Palgrave.

Siep, Ludwig. 1979. *Anerkennung als Prinzip der praktischen Philosophie*. Freiburg/Munich: Karl Alber.

Spiegel, Thomas J. 2022. Normativity between naturalism and phenomenology. *International Journal for Philosophical Studies* 30 (5): 493–518.

Strawson, Peter. 1962. Freedom and resentment. Proceedings of the British Academy, 48: 1–25.

Taylor, Charles. 1992. The politics of recognition. In *Multiculturalism and politics of recognition*, edited by A. Gutmann. Princeton UP: Princeton.

Taylor, Charles. 1985. Human agency and language: Philosophical papers vol. 1. Cambridge: Cambridge.
Toivakainen, Niklas. 2015. The moral roots of conceptual confusion in artificial intelligence research. *American Philosophical Association Newsletters: Philosophy and Computers* 14 (2): 20–31.
Turing, Alan. 1950. Computing machinery and intelligence. *Mind* 59 (236): 433–460.
van den Brink, Bert, and David Owen, eds. 2007. *Recognition and power*. New York: Cambridge University Press.
Walker, Margaret U. 2006. *Moral repair*. Cambridge: Cambridge University Press.
Williams, Robert. 1998. *Hegel's ethics of recognition*. Berkeley: University of California Press.
Williams, Bernard. 1978. Descartes: The project of pure enquiry. New Jersey: Humanities Press.

Towards a Theory of Recognition from Artificial Others

Rosalie Waelen

Abstract The recognitive acts of artificial others do not meet existing criteria for recognition relations between human persons (*interpersonal recognition*). However, even if the recognition received from AI systems (*artificial recognition*) is of a different nature than interpersonal recognition is, practice shows that some kind of acknowledgement is nevertheless experienced in human interactions with artificial intelligence. Given the seemingly inevitable social reality of human-AI relations, it is crucial to theorize the kind of recognition that humans could receive from artificial others. The goal of this chapter is therefore to set the basis for a new, alternative theory of recognition, that captures the nature and moral dimensions of the recognition involved in human-AI interaction. This is done through a discussion of three major conceptual issues: reciprocity, genuineness, and functionality. On the basis of this discussion, a new typology of recognition relations is introduced that distinguishes genuine recognition relations from functional recognition relations. *Genuine recognition* refers to the kind of recognition that can only emerge from the intentional actions of human persons, whose recognitive acts are motivated by genuine recognitive attitudes. *Functional recognition*, on the other hand, focuses on the perception and experience of recipients of recognition, whose self-realization is affected by recognitive acts of other entities, regardless of the intentions or attitudes behind those recognitive acts. It is

R. Waelen (✉)
Institute for Science and Ethics, University of Bonn, Bonn, Germany
e-mail: waelen@iwe.uni-bonn.de

R. Waelen (ed.), *Normative Relations of Recognition in the Age of AI*, Philosophy and Ethics of AI 2, https://doi.org/10.1007/978-3-032-34028-3_11

argued that, while fundamentally different in nature, both categories carry moral weight.

Keywords Human-computer interaction • Human-AI relations • Philosophy of AI • Recognition theory • Self and other • Social philosophy

1 Introduction

Existing theories of recognition do not suffice for making sense of the kind of acknowledgement and confirmation that artificial others can provide. As Chap. 9 by Christina Pinsdorf and Chap. 10 by Arto Laitinen have shown in detail, recognitive acts by AI systems (which I will refer to as *artificial recognition*)[1] do not meet traditional criteria for recognition relations between human persons (i.e., *interpersonal recognition*). As it turns out, artificial recognition is fundamentally different from the recognition received in interpersonal relations. I argue, however, that it is nevertheless important to theorize artificial recognition and the social impact of human-AI relations, because these new types of social relationships are increasingly common. Practice shows that humans relate to AI systems in increasingly intimate ways. Think about people engaging with LLM-based chatbots as if they were their therapists, friends, or even romantic partners. The phenomenon of social relations between humans and artificial others is a new reality, which we have yet to understand fully (Fernandez-Borsot 2025). The aim of this chapter is therefore to set the basis for new theories of recognition as well as new philosophies of technology, that help to understand and critique the moral, social, and psychological implications of social human-AI relations.

Ever since Honneth and others revived Hegel's work on the struggle for recognition in the 1990s, several authors have argued for broadening the scope of recognition, by including non-human lifeforms and other entities into the discussion (Laitinen 2010; Pinsdorf 2016). However, these discussions have focused predominantly on the recipient, i.e., the object of recognition. This chapter widens the scope of recognition even further by reconsidering who can count as a recognizer, i.e., a subject or giver of recognition.[2] Several

[1] I prefer the term 'artificial recognition' to refer to the kind of recognition we could receive from AI systems (or as I like to call them: 'artificial others'), but other terms used for this phenomenon throughout this book are 'algorithmic recognition' (Chap. 2), 'synthetic recognition' (Chap. 3), 'pseudo-recognition' (Chap. 9), and 'simulated recognition' (Chap. 10).

[2] Please note that I do not follow Laitinen (2010) in his distinction between 'recognizing and being recognized', on the one hand, and 'giving and getting recognition', on the other hand. I use 'recognizing',

contributions to this edited volume clarified how interpersonal recognition takes on new dimensions due to the technological mediation of recognition relations and the new ways in which persons can recognize (and misrecognize) each other in the age of AI (see also Waelen and Wieczorek 2022; Waelen 2022). Hence, in terms of giving recognition, there seems to be widespread agreement that interpersonal recognition can now be given *through* technology (adopting the phrase used by Van Uffelen and Secomandi in Chap. 5). This chapter goes one step further, by considering the possibility of receiving recognition *from* technology. I argue that AI not only mediates interpersonal recognition, but can also be a source of recognition itself.

The chapter analyzes three important concepts that form theoretical hurdles to the thesis that AI can be a giver of recognition: 'reciprocity', 'genuineness', and 'functionality'. A separate section is dedicated to each of these three issues (sections 2-4). Part of these discussions are also the concepts of 'moral symmetry' and 'intentionality'. In the final part of the chapter (section 5), I introduce a new typology of recognition relations based on an ontological distinction between *genuine recognition* and *functional recognition*. I argue that asymmetrical and ungenuine recognitive acts are morally relevant because they can nevertheless be functional. Despite not meeting the philosophical criteria for genuine recognition, artificial recognition can potentially still impact people's relation-to-self and the social structures in a society, in ways that interpersonal recognition does.

2 Reciprocity

Honneth, like Fichte and Hegel, argues that relations of recognition need to be reciprocal. However, these thinkers are rather vague with regards to what the condition of reciprocity entails exactly (as also argued by Laitinen 2010). To clarify the matter and distinguish different types of recognition relations, Pinsdorf (2016) proposed to use two separate terms to describe the mutuality involved in recognition: *symmetry* and *reciprocity* (Pinsdorf also discusses this distinction in Chap. 9 of this book). Symmetry entails that both parties—the giver and the recipient of recognition—possess equal moral status. Reciprocity refers to mutually regarding each other as recognizers. What Pinsdorf refers to as reciprocal and non-reciprocal relations of recognition, Ikäheimo and Laitinen (2007) call *dialogical* or *monological* recognition. Dialogical

'giving recognition', and 'being a subject of recognition' interchangeably. The same goes for 'being recognized', 'receiving recognition', and 'being an object of recognition'.

recognition implies mutually regarding each other, while monological recognition only involves recognitive attitudes from the recognizer towards the recognized object.

Since he treats them as interrelated aspects, both symmetry and reciprocity are necessary conditions for recognition according to Honneth (Honneth 1995). Following Hegel's dialectical account of subjectivity, Honneth argues that recognition is all about the experience of knowing oneself through the other. On this account, merely internalizing the perspective of the other does not suffice for self-realization. In our interactions with others, we not only come to understand others as having normative stances towards us, but also reciprocally perceive them in return as objects of our own moral duties. Implied in Honneth's understanding of the relation between self and alterity is the view that recognition is not only reciprocal, but necessarily symmetrical. He holds that people can only perceive themselves as a certain type of person or holder of moral status, when they also understand their partners in interaction as such.

Relevant in this context is Hegel's thesis that the self-realization of the master cannot be successful or meaningful when the slave's recognition of the master is forced by the master. After all, the slave has no choice but to perform recognitive acts towards the master and, moreover, the master does not recognize the slave in return as a person with equal moral status. This argument not only suggests that recognition needs to be reciprocal and symmetrical, but it also seems to imply that recognitive acts need to be genuine. Forced acts can never be genuine ones. I discuss this additional aspect of recognition relations in more detail in the next section (3) on the problem of genuineness.

In contrast to Hegel and Honneth, Pinsdorf argues that *a*symmetrical and *non*-reciprocal recognition relations are possible too (see Pinsdorf 2016 as well as Chap. 9 in this volume). She argues, first of all, that expecting reciprocity denies the moral relations that exist between adults actually capable of reason and morality, on the one hand, and less capable human persons such as children or cognitively impaired adults, on the other hand. According to Pinsdorf, the example of the parent-child bond resembles a symmetrical but non-reciprocal relation of recognition. She reasons that children have equal moral status to adults, due to their inherent human worth, but do not have the same cognitive capacities as adults, which implies that children cannot be expected to reciprocate recognition and carry out moral duties in ways similar to adults. Asymmetrical relations of recognition, secondly, are the kind of moral relations that exist between human persons and non-human lifeforms. Since non-human lifeforms cannot reciprocate, such relations are both asymmetrical and non-reciprocal. Asymmetrical but reciprocal relations, finally, are not included in Pinsdorf's typology of recognition relations.

Reciprocity and symmetry are both necessary conditions according to Honneth, because he conflates the concepts when he refers to "reciprocal recognition" (Honneth 1995). However, he does acknowledge that children or cognitively impaired persons can take part in recognition relations. In contrast to Pinsdorf, Honneth understands the parent-child bond as being both symmetrical and reciprocal. In fact, the parent-child relationship is the paradigmatic case for Honneth's first sphere of recognition: the sphere of love.[3] According to Honneth, reciprocating recognition should be understood in a loose sense; not as a static and equal returning of the same thing, but rather as something that can take on different shapes and forms. While Pinsdorf argues that children or cognitively impaired adults cannot reciprocate recognition, Honneth suggests that they can, but that the recognition they return to their recognizers is of a different nature. For instance, in relations of love between a parent and a child, the child who receives recognition in the form of care reciprocates love in the form of trust and emotional attachment. In the case of legal recognition, people respect other members of society by obeying the law that (ideally) recognizes all members of society as equals through the rights they are granted. In relations of esteem, finally, reciprocity is not necessarily an exchange of the same thing. I can receive esteem for my academic contributions, while I esteem others for their artistic abilities, athletic accomplishments, political activism, and so on. Hence, for Honneth, reciprocal recognition implies that recognition is a two-way street, but the way the subject of recognition relates to the object does not need to be the same as the relation of the object to the subject.

Through her distinction between symmetry and reciprocity, Pinsdorf (2016) extended the scope of recognition beyond interpersonal recognition, to include non-human lifeforms as potential objects of recognition. According to the Hegelian tradition, a self can only come to understand itself as such—be conscious of itself—through interactions with other selves, i.e., other human persons. Most recognition theorists therefore only consider human persons or collectives of persons as possible givers and recipients of recognition.[4] Pinsdorf, however, argues that non-human lifeforms can receive recognition too, or at least to a certain degree. Non-human lifeforms can depend

[3] See Chap. 7 by Wieczorek for an elaborate discussion of recognition involving children.

[4] Gertz (2018) argues that it is wrong to understand Hegel's demand for mutuality in a strict sense, proposing instead that it is enough to meet the condition of mutuality when the other in a recognition relation *appears* to be a conscious self, too. Gertz's interpretation of Hegel opens the possibility for recognition relations with artificial others.

on humans for flourishing, which creates moral relations between humans and other lifeforms. Considering the scope of recognition, Laitinen (2010) also questions who could be possible recipients of recognition and what kind of responses to which kind of features count as recognition. He argues that a definition or theory of recognition that is too strict may rule out important instances of recognition. While Pinsdorf (2016) and Laitinen (2010) have extended the scope of recognition beyond Hegel's and Honneth's strict criteria for such relations, their extensions mostly apply to the recipient's side of the story, i.e., the object of recognition. Who or what can be a giver of recognition has received little reflection so far.

If we consider both symmetry and reciprocity to be necessary conditions for recognition, as Honneth does, only human persons could be givers of recognition. Similarly, from Pinsdorf's Kantian perspective, givers of recognition need to be human persons actually capable of reason and morality, because only such persons have moral autonomy and are thus capable of having moral duties towards others. Pinsdorf accepts asymmetry in recognition relations, but only if it is the receiving party that has a lesser moral status. If we follow these lines of reasoning, artificial others like AI systems could never be givers of recognition, because they will never have the same moral status as the humans who would be the objects of their recognition. However, while I agree that AI does not have equal moral status to human persons, I argue that they can still be givers or subjects of recognition. In other words, I propose that recognition relations with artificial others are necessarily asymmetrical, but can still be reciprocal.

In my view, whether a recognition relation is reciprocal or not is dependent only on the recipient of recognition. I adopt Pinsdorf's understanding of 'reciprocity' as separate from 'moral symmetry', but agree with Honneth that recognition relations are *necessarily* reciprocal. Hence, I accept the thesis that, in a relation where A recognizes B, the two parties do not need to have equal moral status. However, I believe that the entity that receives recognition (B the object) needs to acknowledge its relation to the recognizer (A the subject). The regard that the recipient of recognition (B) returns to its recognizer (A) need not be equal to the recognition received, but some form of awareness and acknowledgement of the received recognition is required. So, I argue, whether A successfully recognizes B depends not on A's possession of reason, morality, or other ontological features, but only on B's judgement of A's acts.

As mentioned above, Honneth perceives relations of recognition between children and parents as reciprocal, even if the moral duties of the child towards the adult are not equal to the duties of the adult towards the child. An adult can care for a child by ensuring their safety, educating them, or making sure they are fed. While a child cannot recognize an adult in the same ways, they can acknowledge the recognition they receive from adult caregivers in the form of affection and attachment. Furthermore, if we accept the thesis that recognition relations can be asymmetrical, we can argue in analogy to adult-child relations that non-human lifeforms like pets or plants can be recipients of recognition. Pets or plants can display having received human care by displaying a form of attachment or flourishing. Just like people's moral duties towards other lifeforms need to be understood in a gradual manner (Pinsdorf 2016), the degree to which non-human lifeforms are able to reciprocate recognition has different degrees too. As an artificial entity, AI does not flourish or grow emotionally attached to other things or beings. Therefore, AI cannot reciprocate received recognition. But that does not mean that AI could not give recognition and be the object of a human's reciprocity. When I receive recognition from an artificial other, I can reciprocate this recognition in the sense that I acknowledge AI's capacity for recognizing me. In my view, acknowledging AI's status as potential giver of recognition does not imply that I simultaneously regard it as a potential object (i.e. receiver) of recognition. Put differently, I propose that reciprocity in human-AI relations implies that humans recognize an AI system as an entity capable of performing recognitive acts. But, unlike Gertz (2018) and Cappuccio et al. (2020), I suggest that AI should not be the other to whom we as humans give our love, respect, or social esteem.

In short, I have argued that *relations of recognition with AI systems are necessarily asymmetrical but potentially reciprocal.* While AI systems do not have equal moral status to human persons and are not able to receive recognition, we can nevertheless regard them as entities capable of giving recognition to humans (or other lifeforms that can reciprocate recognition) in normatively significant ways. As a result, I conclude, we can experience our interactions with artificial others as recognition relations. I will further develop this thesis in the next sections.

3 Genuineness

Many authors in this volume have pointed out that the kind of recognition that AI systems like LLM-based chatbots can give, lacks the moral motivation, empathy, sincerity, or other mental states that characterizes genuine recognition given by human persons. This broadly shared critique of artificial recognition resonates with Ikäheimo and Laitinen's important distinction between recognitive *acts* and recognitive *attitudes:*

> actions are something that we understand as effects, externalizations, or embodiments of intentions or attitudes of persons. A special case of action is action that is performed *in order to express* one's intentions or attitudes to other persons. Concentrating on the recognitive attitudes, we may mean by an 'act of recognition' in a broad sense any act or action that in one way or another is motivated by them. But in a more limited sense, only those actions of A that she performs *in order to express* her love, respect, or esteem towards B, are 'acts of recognition'. We believe that it is important not to identify the immense variety of social actions or practices, in which attitudes of recognition have some motivating role, with 'acts of recognition' in this limited sense. (Ikäheimo and Laitinen 2007, p. 44).[5]

People can have recognitive attitudes without acting upon it and perform recognitive acts without having corresponding recognitive mental states. Without corresponding recognitive attitudes, recognitive acts should be deemed *ungenuine* or *inauthentic*.[6] As Ikäheimo and Laitinen note: "What may seem to be an 'act of recognition' in the limited sense is not a genuine act of recognition (at least in the limited sense) if it is not a sincere attempt to communicate one's true attitudes" (2007, p. 45).

There can be all kinds of reasons for purposefully performing ungenuine recognitive acts. An example of an ungenuine recognitive act is bowing to show respect, which may be done out of cultural habit rather than genuine respect (Ikäheimo and Laitinen 2007). Ungenuine recognition can also be used as a means to an end: A might compliment B, not because A esteems B, but because A needs a favor from B and giving B compliments increases the likelihood that B will agree to help A. While the former example of inauthentic recognition merely has to do with cultural practice, the latter shows how recognition acts can be used instrumentally and strategically. When performed intentionally, ungenuine recognitive acts can be used as an ideological tool to exercise social power. Discussing such ideological dimensions of recognition, Honneth (2007) introduces the term "*false recognition*". False recognition refers to acts of recognition that are ungenuine but intentional. Given recognition's significant role in the fabric of social life and its constitutive effects on a person's relation-to-self, recognitive acts hold significant social power. Hence, by performing recognition acts, for instance through cultural practices around recognition, oppressive social power structures can be kept in place. It is therefore important that social critique not only considers

[5] Ikäheimo further discusses this distinction in his book *Recognition and the Human Lifeform. Beyond Identity and Difference* (2022, pp. 18–22).

[6] In this chapter, I will use the terms 'genuine' and 'authentic' as interchangeable.

instances of genuine recognition (a moral requirement) and misrecognition (a moral problem), but also aims to overcome ideological recognition practices (Honneth 2007).

While this chapter investigates the possibility that AI systems are independent recognizers in a recognition relation, AI can of course also represent or mediated interpersonal recognition. The latter, namely technologically mediated recognition between persons (recognition-through-technology, as Van Uffelen and Secomandi call it in Chap. 5), could be authentic or genuine. Technologically mediated recognition between individual persons or groups can take at least two forms. First of all, technologies can be used as a tool to help humans express recognition. Communication technologies are a fitting example here. From phones to social media platforms, technology enables people to express recognition in new ways and new situations. Communication technologies can overcome communicative hurdles like language barriers or spatial distance. While they also create new possibilities for misrecognition or ideological false recognition, communication technologies have extended the scope of recognition acts (see also Chap. 2 by Senf and Pedersen). A second way in which technology can be said to mediate interpersonal recognition is through the design or development of the system. Whether intentional or not, the normative views of tech-developers or members of society can creep into a technology through the design process or, in the case of AI and algorithms, through training data. To give an example: A biased facial recognition system that consistently flags colored people as potentially dangerous reflects the misrecognition of members of society, as it is societal views that caused the technological biases (Waelen 2022). In both of these examples of recognition-through-technology, the technological (mis)recognition can be deemed genuine, because it ultimately stems from human persons.

It is for similar reasons that recognition scholars agree that laws and institutions can be givers (subjects) of recognition, even though the majority of these scholars also argues that only human persons can be the subject of recognition. For instance, when Honneth (Honneth 1995) writes that people receive recognition in the form of respect from policies and institutions that protect and promote their rights and opportunities, he does not mean to compare the law to a human person capable of recognizing others. Instead, he suggests that the law represents the general other in a society by institutionalizing societal norms. Institutional respect does not imply that the law actively recognizes members of society, as if the law were an independent moral agent with recognitive attitudes and intentionality. Rather, rights and institutions are mediators of the terms of recognition implicitly or explicitly agreed upon within society. By obeying the law, members of society thereupon agree to the conditions of recognition set by the law. In obeying the law, people respect their fellow citizens who are granted rights and opportunities through the policy. In analogy to legal recognition, artificial recognition can entail that AI systems mediate interpersonal recognition. Unlike the law, however, AI can be attributed independent agency. AI systems may not have human-like intelligence, emotions, or consciousness, but, as most philosophers of technology agree, AI systems do have a kind of agency that other non-living entities do not have (Floridi 2023). AI can therefore perform independent recognitive acts, even though it cannot have corresponding recognitive attitudes.

In summary, while recognition-*through*-technology can be genuine, because this kind of technologically mediated recognition is indirectly motivated by the recognitive attitudes of human persons, recognition-*from*-technology is per definition ungenuine, because artificial others can only perform independent recognitive acts without corresponding recognitive attitudes.

4 Functionality

Although the condition of genuineness is given much moral and argumentative weight in existing recognition theories (e.g. Honneth 1995; Ikäheimo 2022), these existing theories fail to specify why exactly it is so important that recognition is genuine. I concluded in the last section that artificial recognition (i.e. recognition-from-technology) is per definition ungenuine, due to the absence of recognitive attitudes. Thus, to make sense of the moral significance of recognitive acts from artificial others, we need to question the long-standing assumption that recognition ought to be genuine.

I have argued above that a recognition relation where A recognizes B, is reciprocal if the recognized object (B) recognizes the recognizing subject (A) as an entity capable of granting recognition. This simultaneously implies that the receiving object (B) is aware of the recognition and impacted by the recognition. What, then, does it take to reciprocally recognize the actions of others as instances of recognition? According to Ikäheimo and Laitinen (2007), adequately understanding recognition acts requires having knowledge of the recognitive attitudes that motivate the recognitive acts:

> Fully understanding any action of a person involves understanding the motives or motivating reasons for the action. The motives or reasons that motivate the actions of persons can tell us of their attitudes towards, in principle, anything that is relevant for the action. (p. 44)

In the context of AI, understanding what motivates the actions of artificial agents implies knowing that their actions are not genuine. To most users, the functioning of AI systems is opaque or even somewhat mysterious, which may lead users to (mistakenly) believe that the recognitive acts from artificial agents are motivated by recognitive attitudes. However, I argue, even those who have a solid understanding of AI's mechanisms could nevertheless accept artificial recognitive acts as a confirmation of their needs, desires, rights, skills, talents, and so on. While philosophers tend to ascribe much weight to the humanity and genuineness of recognitive acts, in practice it might just as well be the case that some people do not care about these philosophical distinctions and norms. Different people could ascribe different values to artificial recognition. In this context, practice-based sociological research is needed to improve our theoretical understanding of human-AI interaction.

A further challenge to the necessity of genuineness is the possibility of *unintentional* genuine recognition. As Ikäheimo and Laitinen (2007) admit, it is possible for a recognitive act to be genuine even if the actor did not intend to express recognition. Take the following example: A might go to their colleague B to ask for help. Subconsciously, A esteems B and that is why A tends to go to colleague B when they need help. A therefore did not intent to express esteem to B when they asked for help, but B may nevertheless experience the request as an act of esteem.[7] In such cases where recognitive acts are unintentional, recognitive attitudes are of an indirect nature, making it even more challenging than usual for recipients of recognition to understand the motivational sources of the recognitive acts of others. In the context of such unintentional-but-genuine recognitive acts, the thesis that successful recognition demands knowing and understanding the motivating reasons behind others' actions implies that B needs to understand A's actions better than A does themselves. In my view, such an epistemic demand seems too strong. Danaher (2019) makes a similar argument in his philosophical discussion of the possibility of robot friendships, arguing that epistemic access may be too high of a demand for friendship—one that we do not always uphold for human-human friendship either.

In practice, human action and moral psychology are complex matters. Ikäheimo and Laitinen (2007) somewhat acknowledge this, writing that "[u]nderstanding the attitudes of others is always fallible, and we can easily be seriously mistaken about the recognitive attitudes of others towards ourselves – say, by interpreting sincere praise as sarcasm or sarcasm as praise." (p. 46) However, despite acknowledging the challenges to their demands for

[7] This case is based on an example discussed by Ikäheimo and Laitinen (2007).

genuine recognition, Ikäheimo and Laitinen hold on to the thesis that successful recognition not only requires genuine attitudes on A's side, but also B's understanding of A's attitudes:

> for genuine recognition to take place, B cannot be seriously mistaken about A's real attitudes towards her. If B just imagines that A has recognitive attitudes towards B, no genuine understanding takes place and we do not have a case of recognition, but, at most, *imagined recognition.* (p. 46)

Although they do not discuss the case of recognition from AI or other technological systems, it is safe to assume that Ikäheimo and Laitinen (2007) would have also ascribed the label 'imagined recognition' to experiences of recognition in human-AI relations. In his contribution to this volume, Laitinen indeed describes recognition by robots as "simulated recognition" and concludes that such simulated recognition does not qualify as recognition in "the relevant sense" (see Chap. 10). Similarly, Pinsdorf (Chap. 9) has argued that AI's recognitive acts are "pseudo-recognition", and Fröding and Peterson (2020) argue that social AI can only "mimic" friendliness and friendship.

Much of the previous literature on recognition emphasizes the importance of genuineness. Admittedly, one can easily see how recognition would have a stronger impact when a person can grasp that the other's recognitive acts are motivated by genuine recognitive attitudes. However, in my view, the powerful impact of genuineness does not take away the possibility that ungenuine acts of recognition can have real and morally relevant consequences too. Genuine human recognition may be morally superior, but rejecting any other forms of recognition means denying the reality and relevance of people's lived experiences in the age of AI. I believe that it is important to address this shortcoming of traditional approaches to recognition, which is why I propose a new conceptual distinction between two types of recognition: *genuine recognition* and *functional recognition.*

The conceptual distinction between genuine and functional recognition represents two different views on the moral relevance of recognition. Recognition theories that emphasize the necessity of genuineness focus on the *source* of an action. Such recognition theories are akin to (but not the same as) deontological or virtue-based moral theories, which place moral weight on the causes of an action and the qualities of the moral agent. Deontological moral theories hold that moral actions are those that are performed out of moral duty. Virtue ethics stresses the importance of developing a virtuous character, which will ensure that one acts rightfully. In contrast, functional recognition theories put more emphasis on the *outcome* of an action. A functional theory of recognition focuses on the impact of recognitive acts on the recipients of recognition, regardless of the nature or source of those recognitive acts. Such functionalist approaches to recognition are then akin to consequentialist moral theories, which take the consequences of actions as morally relevant. While only some recognition relations can be deemed genuine, all recognition relations are by definition functional. After all, if B does not experience A's act of recognition as such, i.e., if B does not receive A's recognition, we cannot speak of a relation of recognition between A and B.

The main benefit of a consequentialist, function-oriented perspective on recognition is that it allows us to make sense of experiences of (mis)recognition in human-AI interaction. The underlying thesis of functional approaches to recognition is that artificial others can function as moral actors because their actions can have morally significant consequences, even if those artificial others are inherently incapable of intentional and genuine moral actions, and do not possess a self that is mutually shaped through interactions with others. Recognition is considered to be important, on this view, not simply because it is the morally right thing to do, but because *experiencing* (mis)recognition affects a person's self-development. The relevance of recognition, under the functional approach, stems from its moral and psychological implications. In other words, it is the experience of the recipient in the recognition relation that is prioritized.

Finally, there are some additional benefits to a functional perspective on the moral importance of recognition. First of all, recognition theories are relational theories that understand the self as shaped through its relationship with others. If the actions of an AI systems can affect a person's sense of self—as the functional approach to recognition suggests—than AI can indeed be a kind of alterity, or "other", even if the artificial other does not possess a self that can mutually be shaped through a recognition relation. Secondly, the functionalist perspective on recognition not only offers a framework to make sense of experiences of recognition in human-technology relations, but also in other types of asymmetric relations—such as the relation between humans and their animal pets. Thirdly, by not dismissing ungenuine acts of recognition, the functionalist approach can be used to address the moral and psychological relevance of the kind of false, ideological practices of recognition theorized by Honneth (2007).

5 Conclusion

The goal of this chapter was to defend the position that the effects of AI's recognition on a person's self-development can be real, even if the source recognition is artificial. In my interpretation of Honneth's work on recognition, taking the experiences of humans in their interactions with AI seriously is ultimately more important than the ontological status of the *other* in the recognition relation. Honneth argues in line with the Frankfurt School tradition of critical theory that social critique should stem from experiences of injustice, i.e., experiences of misrecognition. If interactions with AI are experienced as misrecognizing someone's moral worth, those experiences are a relevant topic for social critique. That is why existing critical theories, including recognition theories, need to be adapted and revised in order to make sense of the new social reality. In the age of AI, humans build social relations with artificial others and have social experiences in their interactions with technological systems. To adjust recognition theory to the age of AI, I have introduced a new typology of recognition relations: *genuine recognition* and *functional recognition*.

My distinction between genuine recognition and functional recognition is the result of a conceptual analysis of reciprocity, moral symmetry, genuineness, intentionality, and functionality. Drawing on Pinsdorf's novel distinction between reciprocity and moral symmetry (Pinsdorf 2016), I first proposed that moral symmetry is *not* a necessary condition in a recognition relation. The asymmetry between human persons and AI agents therefore does not pose an obstacle to theorizing recognition relations with artificial others. I then argued (in contrast to Pinsdorf but akin to Honneth) that reciprocity *is* a necessary condition for recognition, because recognition cannot be deemed relational without reciprocity. Furthermore, I defined reciprocity in a non-intersubjective way, as the recipient's ability to regard the giver of recognition as an entity capable of doing so. Under this understanding, the condition of reciprocity does not pose a hurdle to theorizing AI as a potential giver of recognition either. Next, I discussed the issue of genuineness, using the insightful distinction between recognitive acts and recognition attitudes, introduced by Ikäheimo and Laitinen (2007). On the basis of this distinction, genuine recognition can be defined as recognitive acts that are motivated by sincere recognitive attitudes. To say that an act of recognition is genuine is not the same as saying that it is intentional. Acts of recognition can occur unintentionally, but be genuine nonetheless. I proposed that, although the distinct value of genuine recognition is undeniable, being genuine is not a necessary condition for recognition. While many recognition scholars have stressed the importance of genuineness, none have offered convincing grounds for ruling out the normative implications of ungenuine recognitive acts. Finally, I introduced the concept of functional recognition to refer to cases in which an act of recognition is experienced or perceived as an instance of recognition, irrespective of the intentions, attitudes, or other characteristics of the recognizing party. Functionality focuses on the impact of recognitive acts on the objects of those acts.

Genuine recognition is to be understood as an action- or agent-focused moral theory, whereas functional approaches to recognition are comparable to consequentialist moral theories. I propose treating genuine recognition and functional recognition as complementary concepts—with genuine recognition referring to our moral expectations for successful interpersonal recognition, and functional recognition serving to make sense of the moral dimensions of social relations with non-human or non-living entities. However, it should be noted that the notion of functional recognition could also be used as an opposing or competing view on the moral importance of interpersonal recognition. While genuine recognition excludes the possibility of artificial recognition per definition, and thus needs a complementary theory to make sense of the moral dimensions of ungenuine recognitive acts, a functional perspective on recognition could explain the moral importance of both interpersonal and artificial recognition relations.

References

Cappuccio, M., Peeters, A. & McDonald, W. (2020). Sympathy for Dolores: Moral Consideration for Robots Based on Virtue and Recognition. *Philosophy & Technology, 33* (1): https://doi.org/10.1007/s13347-019-0341-y.

Danaher, J. 2019. The philosophical case for robot friendship. *Journal of Posthuman Studies* 3 (1): https://doi.org/10.5325/jpoststud.3.1.0005.

Fernandez-Borsot, G. 2025. Artificial intelligence and the future of otherness: What kind of other can an AI be for a human? *AI & SOCIETY* 40. https://doi.org/10.1007/s00146-025-02281-2.

Floridi, L. 2023. AI as *agency without intelligence*: On ChatGPT, large language models, and other generative models. *Philososophy & Technology* 36 (15): https://doi.org/10.1007/s13347-023-00621-y.

Fröding, B., and M. Peterson. 2020. Friendly AI. *Ethics and Information Technology* 23. https://doi.org/10.1007/s10676-020-09556-w.

Honneth, A. 1992/1995. *The struggle for recognition. The moral grammar of social conflicts*. Trans. J. Anderson. MIT Press.

Gertz, N. (2018). Hegel, the Struggle for Recognition, and Robots. *Techné, 22* (2). https://doi.org/10.5840/techne201832080.

Honneth, A. 2007. Recognition as ideology. In *Recognition and power*, edited by B. van den Brink and D. Owen, 323–347. Cambridge: Cambridge University Press.

Ikäheimo, H. 2022. *Recognition and the human life-form: Beyond identity and difference*. Routledge.

Ikäheimo, H., and A. Laitinen. 2007. Analyzing recognition: Identification, acknowledgement, and Recognitive attitudes towards persons. In *Recognition and power*, edited by B. van den Brink and D. Owen, 33–56. Cambridge: Cambridge University Press.

Laitinen, A. 2010. On the scope of 'recognition': The role of adequate regard and mutuality. In *Philosophy of recognition: Historical and contemporary perspectives*, edited by H. C. Schmidt am Busch and C. Zurn, 319–342. Rowman & Littlefield.

Pinsdorf, C. 2016. *Lebensformen und Anerkennungsverhältnisse: Zur Ethik der belebten Natur*. De Gruyter.

Waelen, R. 2022. The struggle for recognition in the age of facial recognition technology. *AI and Ethics* 3 (1): https://doi.org/10.1007/s43681-022-00146-8.

Waelen, R., and M. Wieczorek. 2022. The struggle for AI's recognition: Understanding the normative implications of gender bias in AI with Honneth's theory of recognition. *Philosophy & Technology* 35 (53): https://doi.org/10.1007/s13347-022-00548-w.

Index

R. Waelen (ed.), *Normative Relations of Recognition in the Age of AI*, Philosophy and Ethics of AI 2, https://doi.org/10.1007/978-3-032-34028-3